Theoretical Tools for Spin Models in Magnetic Systems

Theoretical Tools for Spin Models in Magnetic Systems

Antonio Sergio Teixeira Pires
Department of Physics, Universidade Federal de Minas Gerais, Belo Horizonte, Brazil

IOP Publishing, Bristol, UK

ISBN 978-0-7503-3879-0 (ebook)
ISBN 978-0-7503-3877-6 (print)
ISBN 978-0-7503-3880-6 (myPrint)
ISBN 978-0-7503-3878-3 (mobi)

DOI 10.1088/978-0-7503-3879-0

Version: 20210401

IOP ebooks

British Library Cataloguing-in-Publication Data: A catalogue record for this book is available from the British Library.

Published by IOP Publishing, wholly owned by The Institute of Physics, London

IOP Publishing, Temple Circus, Temple Way, Bristol, BS1 6HG, UK

US Office: IOP Publishing, Inc., 190 North Independence Mall West, Suite 601, Philadelphia, PA 19106, USA

To Rosangela, Henrique and Guilherme

Contents

Preface

Magnetism is an exciting area of research in condensed matter presenting a variety of exotic phases and critical phenomena and several technological applications. We can study simple models, which are theoretically tractable and yet display quantitative correspondence with real systems.

This book is addressed to people with an interest in magnetism, mainly spin systems. There are excellent advanced textbooks and review papers on the subject. Still, it was my experience that a book was needed, dedicated to the basic concepts with a more elementary treatment for researchers starting in the area. Very little in the text is my original contribution. Most of the material was collected from the literature. However, it would take a long time for the interested student to gather all the topics in the literature. I have tried to make the material mostly self-contained, and references to more advanced treatment are indicated. It is supposed that the reader has some familiarity with the subject at a level as the one presented in the book by Blundel [1]. To keep the text short, I have left out some themes such as one-dimensional models, path-integral formalism and the treatment of continuous systems, and in this context, the nonlinear sigma model.

In chapter 1, I introduce the Heisenberg model and discuss the ground state of the ferromagnet and the antiferromagnet. A short introduction to the Hubbard model is given. In chapter 2, the spin-wave formalism is presented for the ferromagnet and the antiferromagnet. The XY model is studied using a self-consistent harmonic approximation. I finish the chapter with a brief introduction to the Jordan–Wigner transformation and Majorana fermions. In chapter 3, I look at the antiferromagnetic triangular lattice and the antiferromagnetic square lattice in the presence of an external magnetic field. I introduce the Dzyaloshinskii–Moriya interaction, treat the nonlinear spin-wave theory in a mean-field approach, and finally present a modified spin-wave method.

In chapter 4, I describe lattices with three different sites: the honeycomb, the checkerboard, and the Union Jack lattices. To treat the antiferromagnetic lattices, I introduce a generalized Bogoliubov transformation. Chapter 5 is dedicated to the study of formalisms that are more general than the spin-wave theory and can be used to study magnetically disordered phases. I develop the Schwinger boson technique and use it to study valence bond solids and frustrated systems. The use of fermion operators is also mentioned. In chapter 6, I present bond operators adequate to treat dimerized phases and the Schwinger SU(3) bosons. I survey quantum phase transitions, nematic phases, and the Heisenberg model with biquadratic interaction. Chapter 7 is dedicated to the study of dynamics. I introduce the linear response theory, correlation functions, spin transport, the Kubo formulas, the Green function technique, and the memory function formalism. In chapter 8, I develop perturbation theory to calculate the retarded Green function, also called the Feynman propagator, for a many-body system. There are excellent books leading with the subject and discussing in full detail Green functions in the context of perturbation theory. In this chapter, I go straight to the point to teach the reader how to calculate

Feynman diagrams. Magnon topological insulators have been the cause of substantial interest in recent years, and in chapter 9 I explore briefly topological effects in ferromagnets and antiferromagnets. In chapter 10 I present a brief introduction to topological spin liquids. This chapter complements chapter 5, but it is more specific and can be skipped if the reader is not interested in the subject. An overview of numerical methods used in spin models is presented in chapter 11. In appendix A, a short account of group theory is presented to introduce some concepts used in the book. Finally, in appendix B, I finish with a Green function theory specific to spin Hamiltonians. The exercises vary in difficulty and complement the text with specific examples. References are given for the most difficult exercises.

The book by Eriksson *et al* [2] covers some material lacking in the present book, and in that sense both books complement each other.

References

[1] Blundell S 2003 *Magnetism in Condensed Matter* (Oxford: Oxford University Press)

[2] Eriksson O, Bergman A, Bergqvist L and Hellsvik J 2016 *Atomistic Spin Dynamics: Foundations and Applications* (Oxford: Oxford University Press)

Acknowledgments

Chapter 11 was co-authored with Prof. Bismark Vaz da Costa, who is an expert on numerical simulations. I thank him for his help.

I had interactions and collaborations with many colleagues and students who shared insights that appear in this book. I am grateful to two anonymous referees for their suggestions.

I would like to acknowledge the Conselho Nacional de Pesquisa (CNPQ) for partial financial support.

It is a pleasure to acknowledge Emily Tapp, Robert Trevelyan, Chris Benson and others in the editorial and production teams of the IOP Publishing for all their support and help.

I am grateful to all who gave permission to reproduce images included in the figures.

Author biography

Antonio Pires

Antonio Pires graduated from the University of California in Santa Barbara in 1976. He is a professor of Physics at Universidade Federal de Minas Gerais, Brazil, doing research in the area of magnetism. He is a member of the Brazilian Academy of Science. He was the Editor of the Brazilian Journal of Physics and a member of the Advisory Board of the Journal of Condensed Matter Physics. He has published the books *AdS/CFT Correspondence in Condensed Matter* (https://iopscience.iop.org/book/978-1-627-05309-9) and *A Brief Introduction to Topology and Differential Geometry in Condensed Matter Physics* (https://iopscience.iop.org/book/978-1-64327-374-7).

IOP Publishing

Theoretical Tools for Spin Models in Magnetic Systems

Antonio Sergio Teixeira Pires

Chapter 1

The Heisenberg model

We start with a lattice, in one, two, or three dimensions. There are several kinds of lattices, and sometimes we even consider a continuum model. Now, in each lattice site i, we put a spin $\vec{S}_i$, with $\vec{S}_i^{\,2} = S(S+1)$ where S is an integer or a half odd integer. In most cases, one is interested in $S = ½$ and sometimes in $S = 1$. The classical limit where $S \to \infty$ is also of interest. The most straightforward interaction between two spins is given by $\vec{S}_i \cdot \vec{S}_j$. Using quantum mechanics, we can arrive at this term (see, for instance, Auerbach [1] and Maplis [2]). Still, I am not going to do that here and assume that the reader is familiar with the procedure. The basic Hamiltonian in magnetism is the so-called Heisenberg model. This model describes the interaction between spins in a lattice with N sites where generally, $N \to \infty$. This is a many-body problem that usually cannot be exactly solved, and many numerical techniques solve the equations for a finite number of sites. When we do analytical calculations, most of the time, we treat free particles (called quasiparticles) described by a quadratic Hamiltonian. The interaction between particles is then handled via perturbation theory or using a mean-field approach where a quadratic Hamiltonian replaces the true Hamiltonian of the system with renormalized temperature-dependent parameters. Let us start our journey.

The following Hamiltonian represents the Heisenberg model

$$H = \frac{1}{2}\sum_{ij} J_{ij}\vec{S}_i \cdot \vec{S}_j, \tag{1.1}$$

wherein the sum i and j run over all sites on a lattice, $J_{ij} = J_{ji}$ is symmetric, and the factor ½ corrects the double-counting of the bonds. Spin components on the same site obey the commutation relations

$$\left[S_j^{\alpha}, S_j^{\beta}\right] = i\sum_{\gamma} \varepsilon_{\alpha\beta\gamma} S_j^{\gamma} \qquad (\alpha, \beta, \gamma = x, y, z). \tag{1.2}$$

doi:10.1088/978-0-7503-3879-0ch1

Spins on different sites commute with each other. The coefficients J_{ij} are called exchange constants, and for most of the text, we assume $J_{ij} = J$, and nearest-neighbor interactions. If $J < 0$, the energy is minimized when the spins are parallel to each other, and we have a ferromagnetic model. For $J > 0$, antiparallel orientation is instead favored, and we have an antiferromagnet. Defining

$$S_j^+ = S_j^x + iS_j^y, \quad S_j^- = S_j^x - iS_j^y, \tag{1.3}$$

we can write (1.1) as

$$H = \frac{1}{2}\sum_{ij} J_{ij}\left(\frac{S_i^+S_j^- + S_i^-S_j^+}{2} + S_i^zS_j^z\right). \tag{1.4}$$

At high temperatures, the spins are disordered due to the strong thermal fluctuations, and the expectation value of each spin vanishes: $\langle \vec{S}_i \rangle = 0$. For an operator A, we define

$$\langle A \rangle = tr(\rho A), \tag{1.5}$$

where ρ is the density matrix. For mixed states ρ is given by

$$\rho = \sum_n p_n |n\rangle\langle n|, \tag{1.6}$$

where p_n is the probability of occupation of the eigenstate $|n\rangle$. For a pure state $|\psi\rangle$, the density matrix is given by $\rho = |\psi\rangle\langle\psi|$. In the canonical ensemble, as will be the case treated here, we have

$$\langle A \rangle = \frac{1}{Z}\sum_n e^{-\beta E_n} \langle n|A|n\rangle, \tag{1.7}$$

where $Z = tr(e^{-\beta H})$. It may happen that below some critical temperature, T_c, the spins order magnetically, and on average, they point along some definite direction: $\langle \vec{S}_i \rangle \neq 0$. The existence of T_c depends on the dimensionality, the type of lattice, and the range of the interactions. I will discuss the details later. If there is a T_c below this temperature $\langle \vec{S}_i \rangle$ will increase when we decrease T and will reach a maximum value at zero temperature. The critical temperature is called the Curie temperature in ferromagnets and the Néel temperature T_N in antiferromagnets.

1.1 Ground state for the ferromagnet

For a spin S we have the basis $|S, m\rangle$ with

$$\vec{S}^2|S, m\rangle = S(S + 1)|S, m\rangle, \quad S^z|S, m\rangle = m|S, m\rangle. \tag{1.8}$$

Since we are considering lattices where all spins have the same eigenvalue S we write $|m\rangle$. For a lattice, the basis is spanned by the product states $|m_1\rangle, |m_2\rangle, \ldots, |m_N\rangle$.

On this basis, the last term in (1.4) is diagonal, but the other two terms are not. Defining the total spin $\vec{S} \equiv \sum_i \vec{S}_i$ it is easy to show that

$$[S^x, H] = [S^y, H] = [S^z, H] = 0. \tag{1.9}$$

Then we can find simultaneous eigenstates of H, $\vec{S}^2$ and S^z. We will start with states where m_i has the maximum value S. This means that all spins are aligned in the z-direction. We write

$$|\psi\rangle = |S\rangle_1 |S\rangle_2 \ldots |S\rangle_N, \tag{1.10}$$

where $|S\rangle_i$ means $|m = S\rangle$ for the spin at the site i. Here we take $J_{ij} > 0$, and for the ferromagnet, we write equation (1.4) with a minus sign. Acting with H on (1.10) we get

$$H|\psi\rangle = -\frac{1}{2}\sum_{ij} J_{ij}\left(\frac{1}{2}S_i^+S_j^-|\psi\rangle + \frac{1}{2}S_i^-S_j^+|\psi\rangle + S_i^zS_j^z|\psi\rangle\right), \tag{1.11}$$

but $S_i^+S_j^-|\psi\rangle = S_i^-S_j^+|\psi\rangle = 0$, and so we can write

$$H|\psi\rangle = -\frac{1}{2}\sum_{ij} J_{ij}S_i^zS_j^z|\psi\rangle = -\frac{S^2}{2}\sum_{ij} J_{ij}|\psi\rangle. \tag{1.12}$$

Thus $|\psi\rangle$ is an eigenstate of H. There are $2S_{tot} + 1$ of such states since H is invariant under rotations in the spin space. However, a state $|\phi\rangle = |m_1\rangle\, |m_2\rangle \ldots |m_N\rangle$ is generally not an eigenstate of H. We have

$$\begin{aligned}\langle\phi|H|\phi\rangle &= -\frac{1}{2}\sum_{ij}\left(\frac{1}{2}\langle\phi|S_i^+S_j^-|\phi\rangle + \frac{1}{2}\langle\phi|S_i^-S_j^+|\phi\rangle + \langle\phi|S_i^zS_j^z|\phi\rangle\right)\\ &= -\frac{1}{2}\sum_{i,j} J_{ij}m_im_j.\end{aligned} \tag{1.13}$$

If $J_{ij} \geqslant 0$ for all i and j, then $J_{ij}m_im_j \leqslant J_{ij}S^2$ and we conclude

$$\langle\phi|\, H\, |\phi\rangle \geqslant -\frac{S^2}{2}\sum_{ij} J_{ij}. \tag{1.14}$$

Therefore, all eigenenergies, in particular the ground state energy, are larger than or equal to $-(S^2/N)\sum_{ij} J_{ij}$. However, $|\psi\rangle$ is an eigenstate to this energy and therefore it must be a ground state if $J_{ij} \geqslant 0$ for all i, j. We have found then that the fully polarized state is the ground state.

1.2 Spontaneous broken symmetries

Spontaneous symmetry breaking happens when the Hamiltonian has a symmetry and hence a conserved quantity, but the ground state of the system does not have that symmetry. The Hamiltonian of the Heisenberg model given by equation (1.1) is rotationally symmetric. If we rotate all spins by the same angle, the Hamiltonian does not change. The total spin of the system, as we have seen, commutes with the Hamiltonian and hence is conserved. However, the symmetry is broken spontaneously in the ferromagnet's ground state where a single unique direction has been chosen along which all the spins have lined up, and the ground state is not invariant under rotation. When the symmetry that is broken spontaneously is continuous, such as the rotation of spins discussed above, there are infinitely many ground states. (Note that there are a finite number of ground states when the broken symmetry is discrete, such as a spin inversion in the Ising model, to be discussed later.) In the case of spontaneously broken continuous symmetry, the low-lying excitation with energy $E(\vec{k})$ does not have a gap. That is, $E(\vec{k}) \to 0$ as $\vec{k} \to 0$. This result is known as the Goldstone theorem. More details of the theorem can be found in Auerbach [1].

1.3 Ground state for the antiferromagnet

In most cases, the antiferromagnetic ground state is more complicated than the ferromagnetic ground state, and few rigorous results are known about the ground state of the general Heisenberg model. One case where a rigorous result exists is where one has an antiferromagnet on a bipartite lattice. In such a lattice, all sites are divided into two disjoint subsystems A and B so that $J_{ij} = 0$ if $i,j \in A$ or if $i,j \in B$. Hence, one has interactions only between sites from different lattices. As an example, we have the square lattice shown in figure 1.1.

For a bipartite lattice with $J_{ij} \geqslant 0$ in Hamiltonian (1.4), we might naively assume that the ground state is fully polarized but with the spins in opposite directions for the two sublattices. This state is called the Néel state. It is the true ground state for the classical model, but it is not for the quantum case. For instance, the ground state of a four-site spin ½ antiferromagnetic chain is given by

$$|\uparrow\uparrow\downarrow\downarrow\rangle + |\downarrow\downarrow\uparrow\uparrow\rangle + |\downarrow\uparrow\uparrow\downarrow\rangle + |\uparrow\downarrow\downarrow\uparrow\rangle - 2\,|\uparrow\downarrow\uparrow\downarrow\rangle - 2\,|\downarrow\uparrow\downarrow\uparrow\rangle.$$

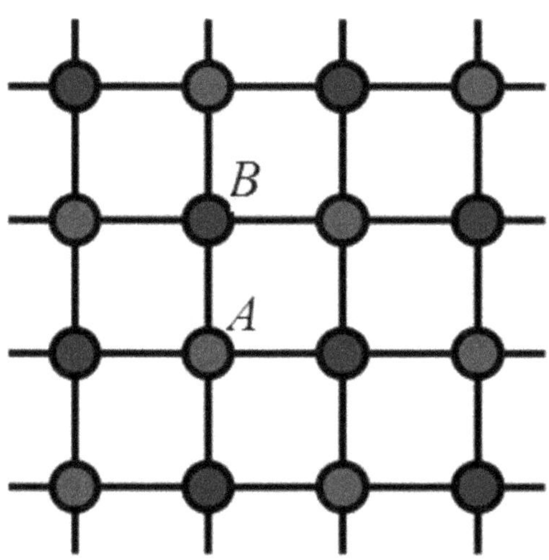

Figure 1.1. The bipartite square lattice with sublattices A and B.

The Néel state can be written as

$$|\psi\rangle = \prod_{i\in A} |S\rangle_i \prod_{j\in B} |-S\rangle_j. \tag{1.15}$$

The states $|\pm S\rangle$ are eigenstates of S^z for the given lattice site with eigenvalues $\pm S$.

We have

$$H\,|\psi\rangle = \sum_{i\in A}\sum_{j\in B} J_{ij}\left(\frac{1}{2}S_i^+S_j^-\,|\psi\rangle + \frac{1}{2}S_i^-S_j^+\,|\psi\rangle + S_i^zS_j^z\,|\psi\rangle\right). \tag{1.16}$$

Note that

$$S_i^+S_j^-\,|\psi\rangle = 0, \tag{1.17}$$

and

$$H\,|\psi\rangle = \sum_{i\in A}\sum_{j\in B} J_{ij} \times \left(\frac{1}{2}[S(S+1) - S(S-1)] \ldots|S-1\rangle_i \ldots|-S+1\rangle_j - S^2\right)|\psi\rangle. \tag{1.18}$$

So, $|\psi\rangle$ is not even an eigenstate of H. We present an important theorem without demonstration. The proof is given in chapter 5 of Auerbach [1].

Marshall's theorem: *The absolute ground state* $|\psi_0\rangle$ *for equal size sublattices A and B is a singlet of the total spin*

$$\mathrm{S}_{\mathrm{tot}}|\psi_0\rangle = 0. \tag{1.19}$$

Marshall's theorem does not uniquely determine the ground state, and many other total-spin singlets are not the ground state.

As we saw, due to the spin rotational invariance of the coupling, the antiferromagnetic interaction drives neighboring spins isotropically antiparallel. The ground state can be achieved by forming an overall singlet with $S_{tot} = 0$ (as follows from Marshall's theorem). The difficulty is that this condition does not fix the state uniquely. The number of singlets grows exponentially as a function of the lattice size (Fazekas [3]), and most of the singlets have a complicated structure. For spin ½ the basic singlet is given by

$$[i, j] = \frac{1}{\sqrt{2}}(|\uparrow_i\downarrow_j\rangle - |\downarrow_i\uparrow_j\rangle). \tag{1.20}$$

Singlet states of more than two spins are obtained by letting all the spins participate in pair bonds. Let us consider as an example a four-site spins ½ Heisenberg model on a square (a single plaquette). The singlet basis can be chosen to consist of the two states shown in figure 1.2.

Other states can be written as a combination of the above basis, as shown in figure 1.3.

The ground state of this four-site Heisenberg model is given by [3]

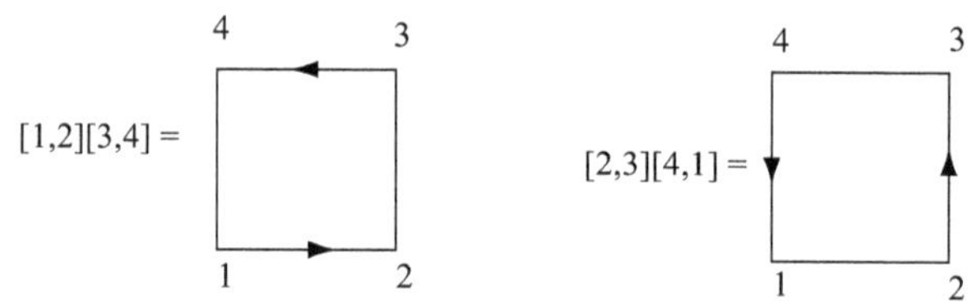

Figure 1.2. Singlet basis.

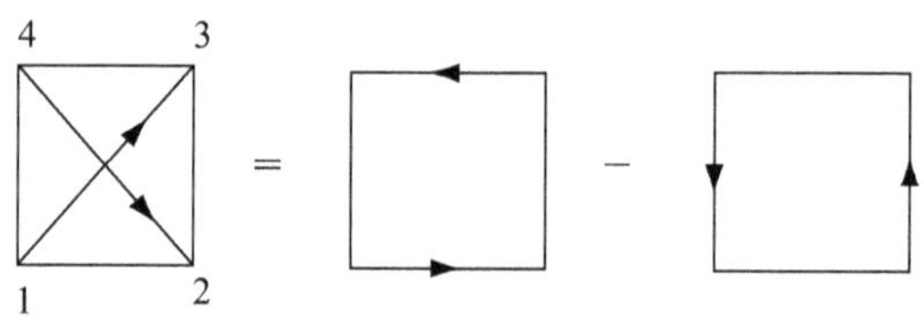

Figure 1.3. [1, 3][4, 2] = [1, 2][3, 4] – [2, 3][4, 1]. This state is a combination of the basis in figure 1.2.

$$\psi_0 = \frac{1}{\sqrt{2}}([1, 2][3, 4] + [2, 3][4, 1]), \tag{1.21}$$

with energy $E_0 = -2J$ (which is quite different from the Néel state). ψ_0 is a superposition of two states with parallel bond configurations on the square. Each state given by the parallel bonds is not an eigenstate of the Hamiltonian, and thus not the ground state. The Hamiltonian flips the bonds on the square between horizontal and vertical configurations of the parallel bonds. The ground state is hence given by the superposition (1.21). The situation remains the same if we go to larger configurations. Only in the thermodynamic limit (when $N \to \infty$), the Néel ground state (for the isotropic square lattice) becomes asymptotically degenerate with the exact singlet ground state. However, there are models for which the ground state is not the Néel state even when the number of sites goes to infinity.

1.4 Excited states for the ferromagnet

In this section, I will consider the excited states of a ferromagnet. I take $S = 1/2$, but it can be easily generalized for other values of spins. For simplicity, let us study a one-dimensional chain. Each spin has two neighbors, and so we can write

$$H = -J \sum_i \vec{S}_i \cdot \vec{S}_{i+1}. \tag{1.22}$$

As we have seen, the ground state of the system $|\psi_0\rangle$ consists of all the spins lying along the z-direction, and we have

$$H|\psi_0\rangle = -\frac{NS^2}{2}|\psi_0\rangle. \tag{1.23}$$

To create an excitation, we flip a spin at site j and consider the state $|\psi_j\rangle = S_j^- |\psi_0\rangle$. This state is the ground state with the spin at site j flipped:

$$|\psi_0\rangle = \ldots \uparrow\uparrow\uparrow\uparrow\uparrow\uparrow\uparrow\uparrow \ldots$$
$$|\psi_j\rangle = \ldots \uparrow\uparrow\uparrow\uparrow\downarrow\uparrow\uparrow\uparrow \ldots$$

Thus, we have changed the total spin of the system by ½ - (- ½) = 1. This excitation has, therefore, an integer spin and is a boson. Acting with H on $|\psi_j\rangle$ we obtain

$$H|\psi_j\rangle = J[(-NS^2 + 2S)\,|\psi_j\rangle - S|\psi_{j+1}\rangle - S|\psi_{j-1}\rangle], \tag{1.24}$$

and we see that $|\psi_j\rangle$ is not an eigenstate of the Hamiltonian. We get an eigenstate by summing over all configurations with a spin flipped at each site j and writing

$$|\psi_q\rangle = \frac{1}{\sqrt{N}}\sum_j e^{iqr_j}|\psi_j\rangle \tag{1.25}$$

where e^{iqr_j} is a phase factor, which describes a plane wave of wave number q. Such waves are known as 'spin waves'. The state $|\psi_q\rangle$ is essentially a flipped spin delocalized. A single spin deviation does not appear on a particular site but is instead distributed over all of the sites in the lattice. It is easy to see that

$$H|\psi_q\rangle = E(q)|\psi_q\rangle, \tag{1.26}$$

where

$$E(q) = -NS^2J + 2JS(1 - \cos qa), \tag{1.27}$$

and a is the lattice spacing. This result can easily be extended to higher dimensions. A ferromagnet is perfectly ordered at $T = 0$, but at non-zero temperature, the order is disrupted by spin waves. The energy of a spin-wave is quantized, and the quantum of energy of a spin-wave is called a *magnon*.

1.5 Translational symmetry

I will consider the same problem taking a linear chain in a ring (that is, with periodic boundary condition). The system is invariant if we displace all spins by one (or more) site. Let T be the translation operator, which takes a spin at the site l into the site $l + 1$. We have

$$TS_l^\gamma T^\dagger = S_{l+1}^\gamma \quad (\gamma = x, y, z), \tag{1.28}$$

and

$$T\,|\alpha\rangle = \alpha\,|\alpha\rangle, \tag{1.29}$$

where $|\alpha\rangle$ and α are eigenvectors and eigenvalues of T. If we have N sites, applying T N times we return to the first site. Thus, $\alpha^N = 1$. The solutions are then

$$\alpha = e^{-iK} \quad \text{with} \quad K = (2\pi/N)n, \quad n = 0, 1, \ldots, N - 1.$$

We can write the eigenstates of H as

$$|S^T, S_z^T, K\rangle, \tag{1.30}$$

where

$$\begin{aligned} \vec{S}^2\,|S^T, S_Z^T, K\rangle &= S^T(S^T + 1)|S^T, S_Z^T, K\rangle, \\ S_Z^T\,|S^T, S_Z^T, K\rangle &= S_Z^T\,|S^T, S_Z^T, K\rangle, \\ T\,|S^T, S_Z^T, K\rangle &= e^{-iK}\,|S^T, S_Z^T, K\rangle, \end{aligned} \tag{1.31}$$

with $0 \leqslant S^T \leqslant NS$, $S_z^T = NS, NS - 1, \ldots, -NS$.

We know from quantum mechanics that translational invariance is associated with the conservation of linear momentum. In the following, I consider the state with one spin flipped. We can write an eigenstate of T as

$$|\phi\rangle = c_1\,|- + + + \ldots\rangle + c_2\,|+ - + + \ldots\rangle + \ldots \tag{1.32}$$

$$T\,|\phi\rangle = c_1\,|+ - + + + \ldots\rangle + c_2\,|+ + - + + \ldots\rangle + \ldots \tag{1.33}$$

But $T\,|\phi\rangle = e^{-iK}\,|\phi\rangle$, which leads to

$$e^{-iK}c_2 = c_1, \ldots., \rightarrow e^{-iK}c_n = c_{n+1}, \ldots., \tag{1.34}$$

so, $c_n = e^{-iKn}$ and thus

$$|\phi\rangle = e^{iK}\,|- + + + \ldots\rangle + e^{2iK}\,|+ - + + \ldots\rangle + e^{3iK}\,|+ + - + +\rangle + \ldots \tag{1.35}$$

We have N eigenvalues of T. We also have one energy for each value of K. For instance, for $N = 4$ we have

$$|\phi\rangle = e^{iK}|- + + +\rangle + e^{2iK}|+ - + +\rangle + e^{3iK}|+ + - +\rangle + e^{4iK}|+ + + -\rangle, \tag{1.36}$$

with $K = 0, \pi/2, \pi, 3\pi/2$.

1.6 Two spin waves

The state with two flipped spins is given by

$$|\psi_{ij}\rangle = S_i^- S_j^-\,|\psi_0\rangle. \tag{1.37}$$

For the ferromagnet, this is a state with two magnons. Note that for $S = ½$ there is no state $|\psi_{ii}\rangle$. We can put the first spin flipped at N sites. The second spin flipped at $N-1$ sites. Since a permutation does not change the state, we have $N(N-1)/2$ states. For $S > ½$, we can have two flipping at the same site, and therefore N additional states, leading to a total of $N(N + 1)/2$ states. I am not going to present the calculation of two spin-wave states here. For this calculation, see Feynman [4]. It is important to mention that now we have bound states, and we can look upon the problem as a scattering of two quasiparticles. In the same way, we can treat three or more spin deviations.

1.7 Long-range order

Suppose that we have a rotationally invariant Hamiltonian H_0. We add an ordering field h, which breaks the rotational symmetry, and write

$$H = H_0 - hS_q^z, \tag{1.38}$$

where

$$S_q^z = \sum_i e^{i\vec{q}\cdot\vec{r}_i}S_i^z. \tag{1.39}$$

The magnetization per site is given by

$$m_q(h) = \frac{1}{NZ}tr\left(e^{-\beta H}S_q^z\right). \tag{1.40}$$

We consider here lattices that are symmetric under reflection about the origin. For $h > 0$ we have $m_q(h) > 0$ since the applied field induces the magnetization. We say that the system has a spontaneously broken symmetry if the magnetization is finite in the thermodynamic limit, as we take the field h to zero from above. That is

$$\lim_{h\to 0^+}\lim_{N\to\infty} m_q(h) \neq 0. \tag{1.41}$$

Note that the order of limits matter. We define the correlation functions as

$$S^{\alpha\alpha}(\vec{q}) = \lim_{h\to 0^+}\frac{1}{ZN}tr\left(e^{-\beta H}S_q^\alpha S_{-q}^\alpha\right). \quad \alpha = x, y, z. \tag{1.42}$$

If $h = 0$, the correlations are independent of the direction α.

1.8 Mermin and Wagner's theorem

The theorem says that for the quantum Heisenberg model

$$H = \frac{1}{2}\sum_{i,j} J_{ij}\vec{S}_i\cdot\vec{S}_j, \tag{1.43}$$

with short-range interactions that obey

$$\frac{1}{2N}\sum_{i,j}|J_{ij}||\vec{r}_i - \vec{r}_j| < \infty, \tag{1.44}$$

there can be no true long-range order at finite temperatures in one and two dimensions. The theorem does not apply at $T = 0$. If there is a unique ground state and a gap in the excitation spectrum, the ground state of the Heisenberg model must be disordered. The proof of the above statement is presented in Auerbach [1]. Note that gapless excitations do not imply long-range order. For example, the spin half Heisenberg antiferromagnet in one dimension has gapless excitation but no long-range order at zero temperature.

1.9 The Ising model

In this book, I will study several spin Hamiltonians. Here I will mention only one specific case: the Ising model described by the Hamiltonian

$$H = -\frac{J}{2}\sum_{ij} S_i^z S_j^z. \tag{1.45}$$

It is the simplest model in magnetism since $|m\rangle_1\ |m\rangle_2 \ldots |m\rangle_N$ is an eigenstate of the Hamiltonian. The model has an ordered phase below the critical temperature in three and two dimensions. It is perhaps one of the most studied models, mainly in statistical physics, and it can be applied to several different systems where we have a variable that can take only two values. Its importance is that it is one of few solvable models in one and two dimensions where we can compute thermodynamic quantities exactly. As concerning magnetic systems, for a material to be Ising-like, the ground state of the ion model must be a doublet well separated from the excited states.

1.10 Brillouin zone

Suppose we have a lattice with lattice vectors [5]

$$\vec{R}_n = n_1\vec{a}_1 + n_2\vec{a}_2 + n_3\vec{a}_3, \tag{1.46}$$

where n_i are integers and $\vec{a}_i$ are the primitive vectors. The reciprocal lattice is generated by basis vectors $\vec{b}_1$, $\vec{b}_2$ and $\vec{b}_3$ given by

$$\vec{b}_1 = \frac{2\pi(\vec{a}_2 \times \vec{a}_3)}{\vec{a}_1 \cdot (\vec{a}_2 \times \vec{a}_3)}, \quad \vec{b}_2 = \frac{2\pi(\vec{a}_3 \times \vec{a}_1)}{\vec{a}_2 \cdot (\vec{a}_3 \times \vec{a}_1)}, \quad \vec{b}_3 = \frac{2\pi(\vec{a}_1 \times \vec{a}_2)}{\vec{a}_3 \cdot (\vec{a}_1 \times \vec{a}_2)}. \tag{1.47}$$

The vectors of the reciprocal lattice are then

$$\vec{K}_n = n_1\vec{b}_1 + n_2\vec{b}_2 + n_3\vec{b}_3 \quad (n_i\text{: integer}). \tag{1.48}$$

If we express the vector $\vec{q}$ as

$$\vec{q} = q_1\vec{b}_1 + q_2\vec{b}_2 + q_3\vec{b}_3, \tag{1.49}$$

then the component q_i has the values

$$q_i = \frac{n_i}{N_i}, \tag{1.50}$$

where N_i is the number of unit cells in the $\vec{a}_i$ direction, and the total number of lattice points is equal to $N = N_1N_2N_3$. We have N independent $\vec{q}$ vectors. The first Brillouin zone is the region surrounded by planes bisecting the vectors connecting the origin and neighboring reciprocal lattice points.

In general, we have to integrate a function in the Brillouin zone. To show how this is done, we start with the simple case of one dimension, where we have the well-known result

$$\int_{-b}^{b} f(x)dx = \sum_{i=1}^{N} f(x_i)\Delta x, \tag{1.51}$$

where $\Delta x = 2b/N$. We can write (1.51) as

$$\frac{1}{N}\sum_{i=1}^{N} f(x_i) = \frac{1}{2b}\int_{-b}^{b} f(x)dx. \tag{1.52}$$

If we have a lattice in one dimension of length a, the reciprocal lattice is given by $-\pi/a \leqslant k \leqslant \pi/a$. With $b = \pi/a$, we have

$$\frac{1}{N}\sum_{i=1}^{N} f(k_i) = \frac{a}{2\pi}\int_{-\pi/a}^{\pi/a} f(k)dk. \tag{1.53}$$

This result can easily be generalized to an integral in the Brillouin zone in d-dimensions

$$\frac{1}{N}\sum_{i} f(\vec{k}_i) = \frac{V}{(2\pi)^d}\int_{BZ} d^d\vec{k}, \tag{1.54}$$

where V is the volume of the unitary cell. For the square or cubic lattice, the calculation is straightforward. If the unitary cell is a rectangle with sides a and b, the Brillouin zone is given by: $-\pi/a \leqslant k_x \leqslant \pi/a, -\pi/b \leqslant k_y \leqslant \pi/b$. However, the integral is more elaborated in other kinds of lattices. Note that it is usual to take the lattice parameter a as unity.

1.11 Mean-field approximation for the classical ferromagnetic Heisenberg model

The magnetization is $\vec{m}_i = \mu_B g \vec{S}_i$, and the total magnetization $\vec{M} = \sum_i \vec{m}_i$. In the case of interaction with only an external field $\vec{B}$, the energy of the system is

$$E = \sum_i E_i = -\mu_0 \vec{B} \cdot \sum_i \vec{m}_i = -\mu_0 \mu_B g \vec{B} \cdot \sum_i \vec{S}_i = -\mu_0 \mu_B g S B \sum_i \cos\theta_i, \tag{1.55}$$

where θ_i is the angle between the magnetic field and the orientation of the spin. The partition function is given by

$$Z_1 = \int \sin\theta \, d\theta d\phi \exp(\beta \mu_0 \mu_B g S B \cos\theta) = 4\pi\left[\frac{\sinh r}{r}\right], \tag{1.56}$$

where $r = \beta\mu_0\mu_B g S B$.

From the free energy $F = -k_B T \ln Z_1$, we get the magnetization (in the field direction)

$$M = -\frac{1}{\mu_0}\left(\frac{\partial F}{\partial B}\right) = \mu_B g S N\left[\coth(r) - \frac{1}{r}\right] = \mu_B g S N L(r), \tag{1.57}$$

where $L(r) = \coth(r) - 1/r$ is the Langevin function.

Now we consider the classical ferromagnetic Heisenberg model in the presence of an external field. We have for the energy

$$E = -J\sum_{\langle i,j\rangle}\vec{S}_i\cdot\vec{S}_j - \mu_0\mu_B g\vec{B}\cdot\sum_i\vec{S}_i. \tag{1.58}$$

In the mean-field approximation, we replace the interaction between spins by an interaction with a field generated by the mean orientation of the spins (that is, we replace one of the $\vec{S}_i$ by its average $\langle\vec{S}_i\rangle$). We write then

$$E = -Jz\langle\vec{S}\rangle\cdot\sum_i\vec{S}_i - \mu_0\mu_B g\vec{B}\cdot\sum_i\vec{S}_i = -(Jz\langle\vec{S}\rangle + \mu_0\mu_B g\vec{B})\cdot\sum_i\vec{S}_i, \tag{1.59}$$

where z is the number of nearest neighbors. We assume that the direction of the z-axis is in the direction of the magnetic field. The direction of $\langle\vec{S}\rangle$ is also in the z-axis direction and its modulus is $\langle S\rangle$. Using equation (1.56) we have for the magnetization

$$M = \mu_B g N S L(f), \quad f = \beta S(Jz\langle S\rangle + \mu_0\mu_B g B). \tag{1.60}$$

However, $M = N\mu_B g\langle S\rangle$, and so we can write

$$\langle S\rangle = SL[\beta S(Jz\langle S\rangle + \mu_0\mu_B g B)]. \tag{1.61}$$

We have to solve (1.61) numerically. For $B = 0$ we find a finite solution for $T < T_c$, where T_c is given by

$$T_c = \frac{JzS^2}{3k_B}. \tag{1.62}$$

We can use the mean-field theory described above only to study ordered phases because we assume $\langle\vec{S}_i\rangle \neq 0$. The antiferromagnet can be treated on the same line, noting that now we have two sublattices.

The reader can easily perform the above calculation for the quantum case. Instead of the integral (1.56), one has a discrete sum over the quantum states that can be calculated using the formula

$$a + ax + ax^2 + \ldots + ax^{L-1} = \sum_{j=1}^{L} ax^{j-1} = \frac{a(1 - x^L)}{1 - x}, \tag{1.63}$$

where L is the number of terms in the series, which in this case is $L = 2J + 1$.

1.12 Landau theory for phase transitions

The Landau theory of phase transitions is based on the view that phase transitions have universal properties, in that they depend only on the nature of the order parameter and not on the finer details of the system. In general, the order parameter vanishes in the disordered phase and takes a non-zero value in the ordered phase. However, correlations and fluctuations are ignored in the theory. The central hypothesis is that we may expand the free energy in a power series in the order parameter in the vicinity of the critical point that we take as the magnetization m. Here we consider a ferromagnet [6]. The equilibrium value of m is then the value that minimizes the free energy. For cases where $F(m, T) = F(-m, T)$, the most general expansion for F is

$$F(m, T) = a(T) + \frac{1}{2}b(T)m^2 + \frac{1}{4}c(T)m^4 + \frac{1}{6}d(T)m^6 + \dots \tag{1.64}$$

where we use fractional coefficients just for convenience. We have

$$\frac{\partial F}{\partial m} = 0 = b(t)m + c(T)m^3 + d(T)m^5 + \dots \tag{1.65}$$

First, we consider the case where $c, d, e, \dots > 0$. To obtain a magnetically ordered state, one must suppose that the coefficient of the term m^2 passes through zero at some temperature T_c. We write then

$$b(T) = \gamma(T - T_c), \tag{1.66}$$

near $T = T_c$. If c is positive, nothing new is added by the term d, and this term may then be neglected. The magnetization (for zero applied field) is found from

$$\gamma(T - T_c)m_s + c(T_c)m_s^3 = 0. \tag{1.67}$$

So that either $m_s = 0$ or

$$m_s^2 = (\gamma/(c(T_c))(T_c - T). \tag{1.68}$$

For $T \geqslant T_c$ the only real root of (1.68) is at $m_s = 0$ because γ and c are positive. Thus, T_c is the Curie temperature. For $T < T_c$ the minimum is at

$$m_s = (\gamma/c(T_c))^{1/2}(T_c - T)^{1/2}. \tag{1.69}$$

This phase transition is a second-order phase transition because the magnetization goes continuously to zero at the transition temperature.

The transition is first order (the magnetization drops discontinuously to zero at the transition temperature) if c is negative. We must then retain d and take it as positive to restrain it from going to minus infinite. The equilibrium condition is

$$\gamma(T - T_c)m_s - |c(T_c)|m_s^3 + d(T_c)m_s^5 = 0. \tag{1.70}$$

So that either $m_s = 0$ or

$$\gamma(T - T_c) - |c(T_c)|m_s^2 + d(T_c)m_s^4 = 0. \tag{1.71}$$

The Landau theory, although very useful, is a mean-field theory and cannot correctly describe critical points, where it should be complemented by scaling laws and renormalization group theory. At a critical point, a thermodynamic function generally contains a term regular in t, where $t = T/T_c - 1$, plus a singular part that behaves like a power of t [7]. The power is the critical *exponent*. When the external magnetic field is zero, we define the following exponents as $t \to 0$:

$$\begin{aligned} M &\sim |t|^{\beta} && \text{(Order parameter)}, \\ \chi &\sim |t|^{-\gamma} && \text{(Susceptibility)}, \\ C &\sim |t|^{-\alpha} && \text{(Heat capacity)}, \end{aligned}$$

where ~ means 'singular part is proportional to' (note that β here is not $1/k_BT$). Critical exponents are universal in that they do not depend on the details of the physical system and are the same for several classes of systems. For instance, for the ferromagnet, they depend on the dimensionality of the lattice, the range of interaction, and the dimensionality of the order parameter (actually the symmetry of the order parameter). The mean-field theory gives $\beta = ½$ (independent of the space dimension). The critical exponents calculated using the mean-field theory differ from the true ones in dimensions 1, 2 and 3. The theory is only correct when the system's space dimension is higher than a certain dimension called the *upper critical dimension.*

1.13 The Hubbard model

The simplest model of a solid is to consider atoms as a collection of sites, each with a single level (orbital). The Pauli principle constrains the sites to four configurations: empty, a single electron with spin-up, a single electron with spin-down, and finally occupation by a pair of electrons with opposite spins. In the so-called tight-binding model, electrons can hop between lattice sites without interacting with each other. In its simplest form, electron hopping can only occur between nearest-neighbor sites, and all hopping processes have the same kinetic energy$-t$. The kinetic energy term in the Hamiltonian is a term that destroys an electron of spin σ on site i and creates it on a neighbor site j (or vice-versa). In the Hubbard model [8], we have an additional term that considers the electrons' interaction via a screened Coulomb repulsion. In the Hamiltonian, we have a term which is zero if the site is empty or has only a single electron, but has the value U if electrons doubly occupy the site with opposite spins since the most significant interaction will be for two electrons on the same site. The Hamiltonian is then written as:

$$H = -t \sum_{\langle i,j\rangle\sigma} \left(c_{i\sigma}^{\dagger}c_{j\sigma} + c_{j\sigma}^{\dagger}c_{i\sigma}\right) + U\sum_{i} n_{i\uparrow}n_{j\downarrow} - \mu\sum_{i}(n_{i\uparrow} + n_{i\downarrow}), \tag{1.72}$$

where

$$n_{\sigma i} = c_{\sigma i}^{\dagger}c_{\sigma i}, \tag{1.73}$$

and $c_{\sigma i}$ are fermion operators. The symbol $\langle i, j\rangle$ means that hopping occurs only between two adjacent sites. The last term is a chemical potential that controls the filling. We refer to the situation where there is one electron per site as 'half-filling'. Depending on the value of U, the model describes a metallic or an insulating state.

If the Coulomb repulsion is large, the electrons are less likely to hop to a neighboring site, and they self-localize on a given lattice site. The electrons can virtually hop to another site and back. However, the material is an insulator and will not conduct. Such a material with an odd number of electrons per unit cell that does not conduct because of Coulomb repulsion is called a Mott insulator, as opposed to band insulators, resulting from filled energy bands. If t/U is large, electrons are essentially free. However, if $U/t \gg 1$, the electrons are localized. The kinetic energy term favors two neighboring site-localized electrons to have antialigned spins, as a hop to a neighboring site is not allowed due to the Pauli exclusion principle. Hence, at half-filling and in the large repulsion limit, the most elementary model that captures the dominant interaction in Mott-insulators is the spin ½ antiferromagnetic Heisenberg model with $J \sim t^2/U$,

$$H = J\sum_{\langle ij\rangle}\vec{S}_i \cdot \vec{S}_j, \tag{1.74}$$

with the spin given by

$$\vec{S}_i = \frac{\hbar}{2}c^{\dagger}_{\sigma i}\tau_{\sigma\sigma'}c_{\sigma' i}, \quad \text{(no sum in } i\text{)} \tag{1.75}$$

where $\vec{\tau}$ are the three Pauli matrices. For the derivation of this result, see Fradkin [9]. Note, however, that equation (1.74) is only the leading term of the effective Hamiltonian. If we go to higher orders, we obtain also ring-exchange processes, second-neighbor interactions, etc. For instance, up to the third order in the strong coupling expansion, we get the following three spin ring-exchange term [10]

$$H_3 = J_3 \sin\phi \sum_{\Delta ijk}\vec{S}_i \cdot (\vec{S}_j \times \vec{S}_k), \tag{1.76}$$

where J_3 is the three spin ring-exchange couplings constant around an elementary triangle $\Delta_{i,j,k}$ on the square lattice formed by the sites i, j, k in a counterclockwise way and ϕ is the magnetic flux enclosed through $\Delta_{i,j,k}$. The scalar spin chirality in $\Delta_{i,j,k}$ is defined by

$$\chi_{ijk} = \vec{S}_i \cdot (\vec{S}_j \times \vec{S}_k). \tag{1.77}$$

This term acts as a fictitious magnetic flux for the Hubble model's conduction electrons and gives rise to a nontrivial topology.

I am not going to study the Hubbard model in this book. The above introduction was only to show that the antiferromagnetic Heisenberg model can be used to study strongly correlated systems.

1.14 Exercises

1.1. Consider the one-dimensional ferromagnetic Ising model with Hamiltonian

$$H = -2J\sum_{i=1}^{N-1} S_i^z S_{i+1}^z - 2JS_N^z S_1^z,$$

where the last term was added so that we have periodic boundary conditions.

a) Show that the operator $\eta_i = 4S_i^z S_{i+1}^z$ has eigenvalues +1 or −1.

b) Using η_i show that the partition function is $Z = [2\cosh(J/2k_BT)]^N$.

1.2. Calculate the partition function and the transition temperature for the quantum ferromagnetic Heisenberg model in three dimensions, using the mean-field approximation.

1.3. Consider the two-site Hubbard model with 2 electrons and $U = 0$. Calculate the ground state eigenstates and eigenenergies.

References

[1] Auerbach A 1998 *Interacting Electrons and Quantum Magnetism* (Berlin: Springer)

[2] Maplis N 2007 *Quantum Theory of Magnetism* (Singapore: World Scientific)

[3] Fazekas P 2003 *Lecture Notes on Electron Correlation and Magnetism* (Singapore: World Scientific)

[4] Feynman R P 1974 *Statistical Mechanics: A Set of Lectures* (Reading, MA: W A Benjamin)

[5] Yosida K 1998 *Theory of Magnetism* (Berlin: Springer)

[6] Plischke M and Bergersen B 2003 *Equilibrium Statistical Physics* (Singapore: World Scientific)

[7] Huang K 2001 *Statistical Physics* (London: Taylor & Francis)

[8] Alet F, Walczak A M and Fisher M P A 2006 Exotic quantum phases and transitions in correlated matter *Physica* A **369** 122–42

[9] Fradkin E 1991 *Field Theories of Condensed Matter Systems* (Redwood City, CA: Addison-Wesley Publishing Company)

[10] Tao L 2019 What does the giant thermal Hall effect observed in the high-temperature superconductors imply? arXiv:1911.03979

IOP Publishing

Theoretical Tools for Spin Models in Magnetic Systems

Antonio Sergio Teixeira Pires

Chapter 2

Spin waves I

The commutation of two spin operators is itself an operator, making it more difficult to work with them directly. It is, therefore, convenient to represent spin operators in terms of bosonic or fermionic operators. There are several ways to do that, as I will show in this book. One of these techniques is the so-called spin-wave theory. This theory refers to any method in which we find magnon dispersion of a ferromagnet or antiferromagnet by studying the fluctuations about its classical ground state. To apply this formalism, we need to know *a priori* the ground state, and it can be used only when long-range magnetic order exists, and we know what kind of order it is.

2.1 Ferromagnet

We start with the ferromagnet described by the Hamiltonian

$$H = -J\sum_{\langle i,j\rangle} \vec{S}_i \cdot \vec{S}_j, \tag{2.1}$$

where $\langle i, j\rangle$ means nearest-neighbor interaction. In chapter 1, I wrote a factor ½ in the Hamiltonian. Here I put the factor ½ when I take the Fourier transform, as usually is done in the literature. As was discussed in chapter 1, the ferromagnet's ground state is exactly known, with all spins aligned along a direction that I will take as the z-direction. As in chapter 1, we have the raising and lowering operators given by $S_l^{\pm} = S_l^x \pm S_l^y$, $n_l = S - S_l^z$. The eigenvalues of n_l are integers going from 0 to $2S$ since the eigenvalues of S_l^z are $m = -S, -S+1,\ldots, S$. We assume S is a fixed value, so the only variable is m. Let $\phi(m)$ be an eigenvector of the eigenvalue m of S_l^z. We have (writing $n = S - m$):

$$\begin{aligned} S_l^- \phi(m) &= \sqrt{(S+m)(S-m+1)}\,\phi(m-1) \\ &= (2S)^{1/2}\sqrt{(n_l+1)\left(1-\frac{n_l}{2S}\right)}\,\phi(m-1). \end{aligned} \tag{2.2}$$

doi:10.1088/978-0-7503-3879-0ch2

We can also write

$$S_l^- \phi(n_l) = \sqrt{2S}(n_l + 1)^{1/2}\left(1 - \frac{n_l}{2S}\right)^{1/2} \phi(n_l + 1). \tag{2.3}$$

Similarly, we obtain

$$S_l^+ \phi(n_l) = \sqrt{2S}\, n_l^{1/2}\left(1 - \frac{n_l - 1}{2S}\right)^{1/2} \phi(n_l - 1). \tag{2.4}$$

Now we introduce boson operators a_l and $a_l^\dagger$ such that $a_l^\dagger a_l = n_l$. Note that these operators obey the commutation relation $[a_l, a_l^\dagger] = 1$. One has then

$$\begin{aligned} a_l^\dagger a_l \phi(n_l) &= n_l \phi(n_l), \\ a_l^\dagger \phi(n_l) &= \sqrt{n_l + 1}\, \phi(n_l + 1), \\ a_l \phi(n_l) &= \sqrt{n_l}\, \phi(n_l - 1). \end{aligned} \tag{2.5}$$

Comparing (2.5) with (2.3) and (2.4) we get

$$\begin{aligned} S_l^- &= \sqrt{2S}\, a_l^\dagger \left(1 - \frac{a_l^\dagger a_l}{2S}\right)^{1/2}, \\ S_l^+ &= \sqrt{2S}\left(1 - \frac{a_l^\dagger a_l}{2S}\right)^{1/2} a_l, \quad S_l^z = S - a_l^\dagger a_l. \end{aligned} \tag{2.6}$$

The representation (2.6) is called the Holstein–Primakoff representation (HP). The factor $\sqrt{1 - n_l/2S}$ in the raising and lowering operators limit the number of bosons one can have at a given site to $2S$. The eigenvalues of $a_l^\dagger a_l$ are integers (0, 1, 2,…), while the eigenvalues of n_l are restricted to 0, 1,…, $2S$. However, the matrix elements of S_l^+ and S_l^- between $2S$ and $2S + 1$ vanish, and hence there is no mixture of states with $n_l > 2S$ with those of $n_l \leqslant 2S$. Note that the boson number must satisfy the condition

$$n_l \leqslant 2S. \tag{2.7}$$

The Hamiltonian written in terms of the representation (2.6) is exact. It can easily be shown that the relation $\vec{S}^2 = S(S + 1)$ and the commutation relations (1.2) follow from the HP representation and the bosonic commutation relations. The vacuum of the a_l bosons (i.e., the state with no such bosons) is the state corresponding to the maximum eigenvalues S of S^z.

We introduce Fourier-transformed boson operators as

$$a_k = \frac{1}{\sqrt{N}} \sum_j e^{-i\vec{k}\cdot\vec{r}_j} a_j, \quad a_k^\dagger = \frac{1}{\sqrt{N}} \sum_j e^{i\vec{k}\cdot\vec{r}_j} a_j^\dagger, \tag{2.8}$$

where N is the number of sites. The inverse transformation is then given by

$$a_j = \frac{1}{\sqrt{N}}\sum_k e^{i\vec{k}\cdot\vec{r}_j}a_k, \quad a_j^\dagger = \frac{1}{\sqrt{N}}\sum_k e^{-i\vec{k}\cdot\vec{r}_j}a_k^\dagger. \tag{2.9}$$

It is easy to show using the relation $(1/N)\sum_j e^{i(\vec{k}-\vec{q})\cdot\vec{r}_j} = \delta_{kq}$, that the commutator satisfies the bosonic commutation relation

$$\left[a_k, a_q^\dagger\right] = \delta_{kq}, \quad [a_k, a_q] = \left[a_k^\dagger, a_q^\dagger\right] = 0. \tag{2.10}$$

The operator $a_k^\dagger$ creates a magnon with a wavevector $\vec{k}$, and the operator a_k destroys a magnon with a wavevector $\vec{k}$. If we consider only nearest-neighbor interactions, we can write $\vec{j} = \vec{i} + \vec{\delta}$, where $\vec{\delta}$ is a vector connecting nearest-neighbor sites. This chapter will treat only the standard lattice in one dimension, the square lattice in two dimensions, and the cubic lattice in three dimensions. Other lattices will be studied later.

The low-energy excitations correspond to the spins' small collective oscillations around the ordering direction (which we chose to be the z-direction). These oscillations make $\langle S_l^z\rangle$ less than the maximum value S. This means that the boson number $\langle n_l\rangle$ is non-zero. If $\langle S_l^z\rangle$ is much smaller than S the reduction in S_l^z is small, and we can expand the square root in the small parameter $\langle n_l\rangle/S$, using $(1 + x)^\alpha \approx 1 + \alpha x$ (with $|\alpha x| \ll 1$). That is, we will be concerned chiefly with low-lying states of the system such that $\langle n_l\rangle/S \ll 1$.

To treat the quadratic Hamiltonian, we take the expressions

$$S_j^+ \approx \sqrt{2S}\,a_j, \quad S_j^- \approx \sqrt{2S}\,a_j^\dagger, \quad S_j^z = S - a_j^\dagger a_j, \tag{2.11}$$

into the Hamiltonian (2.1). The higher-order terms introduce interactions between magnons that we will neglect for now. Keeping only terms quadratic in bosonic operators in the Hamiltonian is referred to as *the linear spin-wave theory*. One can then check if the spin deviations are indeed small. If it is not the case, we have an indication that quantum fluctuations destroy the long-range magnetic order, and the theory cannot be used. Note also that the spin-wave approach is a large-S expansion and cannot capture highly quantum states that are rotationally symmetric such as the valence bond solids to be discussed in chapter 5. Nevertheless, the approach works quite well for ferromagnets, where the classical order is *a priori* known.

The quadratic Hamiltonian in the linear spin-wave theory is then written as

$$H_1 = -JNS^2z/2 - JS\sum_{i,\delta}[a_i a_{i+\delta}^\dagger + a_i^\dagger a_{i+\delta} - a_i^\dagger a_i - a_{i+\delta}^\dagger a_{i+\delta}], \tag{2.12}$$

where z is the number of nearest neighbors and is given by $2d$ where d is the lattice's dimension. Here we have used $\sum_{i,j} = \frac{1}{2}\sum_{i,\delta}$, and now the factor ½ corrects the double-counting of the bonds. Note that some authors call the quadratic Hamiltonian H_0, while others keep the same letter H. I will use H_1. To proceed, I take the Fourier transform. For instance,

$$\begin{aligned}\sum_i\sum_\delta a_i^\dagger a_{i+\delta} &= \frac{1}{N}\sum_i\sum_\delta\sum_k\sum_q e^{-\vec{k}.\vec{r}_i}a_k^\dagger e^{i\vec{q}.(\vec{r}_i+\vec{\delta})}a_q \\ &= \frac{1}{N}\sum_i\sum_\delta\sum_k\sum_q e^{-(\vec{k}-\vec{q})\cdot\vec{r}_i}a_k^\dagger e^{i\vec{q}\cdot\vec{\delta}}a_q = \frac{1}{N}\sum_k\sum_q\sum_\delta N\delta_{kq}e^{i\vec{q}\cdot\vec{\delta}}a_k^\dagger a_q \\ &= \sum_k\sum_\delta e^{i\vec{k}\cdot\vec{\delta}}a_k^\dagger a_k,\end{aligned} \tag{2.13}$$

where I have again used $\sum_i e^{i(\vec{k}-\vec{q})\cdot\vec{r}_i} = N\delta_{kq}$. Defining

$$\gamma_k = \frac{1}{z}\sum_\delta e^{i\vec{k}\cdot\vec{\delta}}, \tag{2.14}$$

we can write the harmonic Hamiltonian as

$$H_1 = E_0 + \sum_k \omega_k a_k^\dagger a_k, \tag{2.15}$$

where

$$E_0 = -JNS^2z/2, \quad \omega_k = SJz(1 - \gamma_k). \tag{2.16}$$

The Hamiltonian (2.15) describes noninteracting spin waves (magnons) with dispersion ω_k. The expression for ω_k agrees with the one calculated in section 1.4. In the presence of an external magnetic field, B, applied in the z-direction we add a term $-\mu_B gB\sum_i S_i^z$ to the Hamiltonian, leading to a term $\mu_B gB$ in ω_k.

We have

$$\gamma_k = \cos k \text{ in 1d}, \quad \gamma_k = \frac{1}{2}(\cos k_x + \cos k_y) \text{ in 2d}. \tag{2.17}$$

In the following, I show the calculation of γ_k for the simple cubic lattice in 3d. The nearest neighbors are along the $\pm x$, $\pm y$ and $\pm$ z-axes at a distance a, which we take as the unity. We have

$$\begin{aligned}\gamma_k &= \frac{1}{6}(e^{ik_x} + e^{-ik_x} + e^{ik_y} + e^{-ik_y} + e^{ik_z} + e^{-ik_z}) \\ &= \frac{1}{3}(\cos k_x + \cos k_y + \cos k_z).\end{aligned} \tag{2.18}$$

Using periodic boundary conditions leads to $a_1 = a_{1+N_x\hat{x}}$ where N_x is the number of sites in the x-direction. This condition is satisfied if $e^{ik_xN_x} = 1$, i. e., $k_x = 2\pi n_x/N_x$ where n_x is an integer. The same procedure applies to the other directions. We note that $\sum_k \gamma_k = 0$ for all the lattices discussed above, as can be checked by a direct calculation.

The long-wavelength limit of the ferromagnetic spin-wave dispersion (2.16) vanishes as

$$\omega_k = SJ\,|\vec{k}|^2. \tag{2.19}$$

This mode is the gapless Goldstone mode mentioned in section 1.2, which is a consequence of the spontaneously broken symmetry of the ferromagnetic ground state.

The lowest-order correction to the ground state magnetization at finite temperatures is given by

$$\Delta m = m(0) - m(T) = S - \frac{1}{N}\langle S_T^z \rangle = \frac{1}{N}\sum_k n_k. \tag{2.20}$$

For free magnons n_k is the Bose–Einstein occupation number

$$n_k = (e^{\beta\omega_k} - 1)^{-1}. \tag{2.21}$$

Hence,

$$\Delta m = \frac{1}{(2\pi)^d}\int \frac{d^dk}{e^{\beta\omega_k} - 1}. \tag{2.22}$$

For $T > 0$ the integral in (2.22) diverges in one and two dimensions. (For small k, $n_k \approx (\beta\omega_k)^{-1} \approx (\beta SJk^2)^{-1}$, $d^dk \propto k^{d-1}dk$. Then we need $d - 1 \geqslant 2 (d \geqslant 3)$ to cancel the divergence in $k = 0$ in equation (2.22).) That is consistent with Mermin and Wagner's theorem presented in section 1.8, which rules out a finite spontaneous magnetization in $d = 1$ and 2 at all non-zero temperatures. For $d = 3$, it can be shown that [1] $\Delta m = 0.117 g\mu_B (k_B T/JS)^{3/2}$ as $T \to 0$. We can write

$$m(T) = m(0)\left[1 - \left(\frac{T}{T_c}\right)^{3/2}\right], \tag{2.23}$$

where the critical temperature T_c is given by

$$T_c = \left(\frac{m(0)}{0.117 g\mu_B}\right)^{3/2} \frac{JS}{k_B}. \tag{2.24}$$

Let us consider now the anisotropic ferromagnet described by the Hamiltonian

$$H = -J\sum_{\langle i,j\rangle}\left(S_i^x S_j^x + S_i^y S_j^y + \lambda S_i^z S_j^z\right), \tag{2.25}$$

with $\lambda > 1$. The rotational symmetry is now broken, but $[S_T^z, H] = 0$. In the ground state, the spins are aligned along the z-direction (now we cannot choose the magnetization direction arbitrarily, as we did before for the isotropic model). The symmetry is broken, but not spontaneously broken. The dispersion relation for the quadratic term is now given by

$$\omega_k = SJz(\lambda - \gamma_k), \tag{2.26}$$

and we have a gap at $k = 0$ given by $\omega_0 = SJZ(\lambda - 1)$. The integral in Δm is convergent for $d = 3$ and $d = 2$ and now we have an ordered state below a critical temperature, even in two dimensions.

Up to now, we have used the Holstein–Primakoff representation, but we can also use a generalized bosonic representation given by [2]:

$$S_i^+ = \sqrt{2S}\left(1 - \frac{a_i^\dagger a_i}{2S}\right)^\alpha a_i,\ S_i^- = \sqrt{2S}\,a_i^\dagger\left(1 - \frac{a_i^\dagger a_i}{2S}\right)^{1-\alpha},\ S_i^z = S - a_i^\dagger a_i, \quad (2.27)$$

where $0 \leqslant \alpha \leqslant 1$. This transformation satisfies the commutation relations $[S^+, S^-] = 2S^z$, $[S^\pm, S^\pm] = \mp S^\pm$ for any α. However, the transformation is non-unitary, i.e., $(S^+)^\dagger \neq S^-$, unless $\alpha = 1/2$, in which case, we recover the Holstein–Primakoff representation. For $\alpha = 1$ we get the so-called Dyson–Maleev representation

$$S_i^+ = \sqrt{2S}\left(1 - \frac{a_i^\dagger a_i}{2S}\right)a_i,\ S_i^- = \sqrt{2S}\,a_i^\dagger. \quad (2.28)$$

In this representation, we get rid of the square-root factor, and the interactions can be expressed by a finite number of quartic boson operators at the expense of the nonhermiticity of the Hamiltonian. However, now the nonphysical part of the Hilbert space (with $n_i > 2S$) becomes connected to the physical part because $S^-|-S\rangle = 0$ is not automatically ensured. In the linear spin-wave limit, the Holstein–Primakoff and the Dyson–Maleev representations give the same results. In the literature, the distinction between dynamical and kinematical interactions is made. The dynamical part is the one that expresses multi-particle processes in the Hamiltonian. The kinematical part results in the condition to include only physical states, i. e., states with at maximum $2S$ spins per site [3].

2.2 Antiferromagnet

In the case of an antiferromagnet, several kinds of ground states are possible depending, for instance, on the lattice, and these ground states are, most of the time, qualitatively different from each other. In general, the Heisenberg antiferromagnet, as was mentioned in chapter 1, is a complex and challenging problem to be solved [4, 5]. In this section, I will consider a square or cubic lattice.

The Néel state is not the true ground state for the quantum model. However, numerical calculations (and comparison with experimental data in three dimensions) have shown that it is a good approximation, at least in a qualitative sense [6]. Then we consider that the spins on sublattice A point predominately in one direction that we will take as the z-direction, and the spins on sublattice B point predominately in the opposite direction.

On sublattice A, we use the standard HP representation

$$
\begin{aligned}
S_i^+ &= \sqrt{2S}\left(1 - \frac{a_i^\dagger a_i}{2S}\right)^{1/2} a_i, \\
S_i^- &= \sqrt{2S}\, a_i^\dagger\left(1 - \frac{a_i^\dagger a_i}{2S}\right)^{1/2}, \\
S_i^z &= S - a_i^\dagger a_i. \quad i \in A.
\end{aligned}
\tag{2.29}
$$

On the B sublattice where the spin projection is in the $-\hat{z}$ direction, we will make the changes $S^+ \leftrightarrow S^-$, $S^z \to -S^z$, and write

$$
\begin{aligned}
S_j^+ &= \sqrt{2S}\, b_j^\dagger\left(1 - \frac{b_j^\dagger b_j}{2S}\right)^{1/2}, \\
S_j^- &= \sqrt{2S}\left(1 - \frac{b_j^\dagger b_j}{2S}\right)^{1/2} b_j, \\
S_j^z &= - S + b_j^\dagger b_j. \quad j \in B.
\end{aligned}
\tag{2.30}
$$

We have introduced boson operators a_i for sublattice A and boson operators b_j for sublattice B (note that these operators are not magnon variables). Expanding the square roots and inserting the HP expressions into the Hamiltonian (2.1) with a positive sign, we get the quadratic Hamiltonian:

$$
\begin{aligned}
H_1 = J\sum_{i \in A}\sum_{\delta}&[S(a_i b_{i+\delta} + a_i^\dagger b_{i+\delta}^\dagger) + S(a_i^\dagger a_i + b_{i+\delta}^\dagger b_{i+\delta}) - S^2] \\
+ J\sum_{j \in B}\sum_{\delta}&\left[S\left(b_j a_{j+\delta} + b_j^\dagger a_{j+\delta}^\dagger\right) + S\left(b_j^\dagger b_j + a_{j+\delta}^\dagger a_{j+\delta}\right) - S^2\right].
\end{aligned}
\tag{2.31}
$$

The Fourier-transformed operators are

$$
a_k = \frac{1}{\sqrt{N_A}}\sum_{i \in A} e^{-i\vec{k}\cdot\vec{r}_i} a_i, \qquad b_k = \frac{1}{\sqrt{N_B}}\sum_{j \in B} e^{-i\vec{k}\cdot\vec{r}_j} b_j, \tag{2.32}
$$

where $N_A = N_B = N/2$ is the number of lattice sites in each sublattice.

The magnetic Brillouin zone is different from the crystal Brillouin zone, as we will see below. Let us first consider the one-dimensional system. If the lattice spacing is 'a' the spacing between neighboring sites in a sublattice is $2a$. So periodic boundary conditions demand $a_{i+N_A 2\hat{x}} = a_i$, which gives $k_x = 2\pi n_x/(2N_A)$. Choosing $n_x = -N_A/2, \ldots, N_A/2 - 1$ gives $k_x \in [\pi/2, \pi/2)$. Thus, the length of the magnetic Brillouin zone is half the length of the crystal Brillouin zone $[\pi, \pi)$. We can now generalize the result to the other directions. In figure 2.1, I show the magnetic BZ for the square lattice in two dimensions.

Using $\sum_{i \in A} e^{i(\vec{q}-\vec{k}).\vec{n}} = (N/2)\delta_{q,k}$, the Fourier-transformed Hamiltonian takes the form (where I have renamed $-\vec{k}$ as $\vec{k}$, where needed)

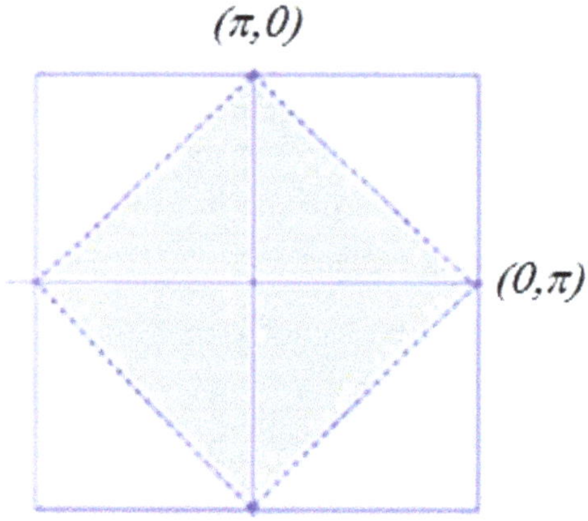

Figure 2.1. The magnetic Brillouin zone for the square lattice antiferromagnet. The magnetic Brillouin zone is a square oriented at a 45-degree angle with respect to the crystal Brillouin zone and has half its area.

$$H_1 = -NJS^2z/2 + JSz\sum_k[\gamma_k(a_kb_{-k} + a_k^\dagger b_{-k}^\dagger) + a_k^\dagger a_k + b_k^\dagger b_k], \tag{2.33}$$

and γ_k was defined in equation (2.14). To diagonalize the Hamiltonian (2.33), we use the so-called Bogoliubov transformation in the form presented in textbooks. In chapter 4, I will give a demonstration of this transformation for more general models. In this procedure, one introduces new creation and annihilation operators $\alpha_k^\dagger$, $\beta_k^\dagger$, α_k, β_k and write

$$\alpha_k = u_k a_k - v_k b_{-k}^\dagger, \quad \beta_k = u_k b_k - v_k a_{-k}^\dagger, \tag{2.34}$$

where u_k and v_k are real functions of $\vec{k}$ which are to be determined. The relation $[\alpha_k, \alpha_q^\dagger] = [\beta_k, \beta_q^\dagger] = \delta_{kq}$ demands

$$u_k^2 - v_k^2 = 1. \tag{2.35}$$

This equation can be satisfied if we write

$$u_k = \cosh\theta_k, \quad v_k = \sinh\theta_k. \tag{2.36}$$

On the other hand, $[\alpha_k, \beta_q] = 0$ leads to $u_k v_{-k} = u_{-k} v_k$, which is satisfied if $u_{-k} = u_k$ and $v_{-k} = v_k$. The inverse relation is given by

$$a_k = u_k\alpha_k + v_k\beta_{-k}^\dagger, \quad b_k = u_k\beta_k + v_k\alpha_{-k}^\dagger. \tag{2.37}$$

Taking (2.37) into (2.33) yields

$$\begin{aligned} H_1 = &- NJS^2z/2 + JSz\sum_k\Big\{(2\gamma_k u_k v_k + u_k^2 + v_k^2)\big(\alpha_k^\dagger\alpha_k + \beta_k^\dagger\beta_k\big) \\ &+ 2(\gamma_k u_k v_k + v_k^2) + [\gamma_k(u_k^2 + v_k^2) + 2u_k v_k]\big(\alpha_k\beta_{-k} + \alpha_k^\dagger\beta_{-k}^\dagger\big)\Big\}. \end{aligned} \tag{2.38}$$

Now we demand that the term which is not diagonal vanishes, that is

$$\gamma_k(u_k^2 + v_k^2) + 2u_k v_k = 0. \tag{2.39}$$

Using the relations $\cosh^2 x + \sinh^2 x = \cosh 2x$, $2\cosh x \sinh x = \sinh 2x$, and equation (2.36), we can write equation (2.39) as

$$\tanh 2\theta_k = -\gamma_k. \tag{2.40}$$

If the lattice has inversion symmetry, as is the case studied here, γ_k is symmetric, i. e., $\gamma_k = \gamma_{-k}$. This condition leads to $\theta_k = \theta_{-k}$. The Hamiltonian can finally be written as

$$H_1 = E_0 + \sum_k \omega_k\left(\alpha_k^\dagger \alpha_k + \beta_k^\dagger \beta_k\right), \tag{2.41}$$

where

$$E_0 = -NJS^2z/2 - NJSz/2 + \sum_k \omega_k, \tag{2.42}$$

and

$$\omega_k = JSz\sqrt{1 - \gamma_k^2}. \tag{2.43}$$

The operators $\alpha_k^\dagger$ and $\beta_k^\dagger$ create magnon excitations with wavevector $\vec{k}$ and energy ω_k. Although we have started with the Néel configuration, the ground state of the quantum antiferromagnet is the vacuum of the α and β bosons (not of the a and b bosons). That is not the Néel state anymore but contains configurations with many spin flips with respect to the Néel state, which reduces the long-range Néel order at $T = 0$.

By inserting the Holstein–Primakoff transformation into S^z we obtain

$$S^z = \sum_i (S_{iA}^z + S_{iB}^z) = \sum_k S_k^z = \sum_k (-a_k^\dagger a_k + b_k^\dagger b_k). \tag{2.44}$$

Using the Bogoliubov transformation, we find

$$S^z = \sum_k \left(-\alpha_k^\dagger \alpha_k + \beta_k^\dagger \beta_k\right), \tag{2.45}$$

thus $\langle 0|\alpha_k S^z \alpha_k^\dagger|0\rangle = -1$ and $\langle 0|\beta_k S^z \beta_k^\dagger|0\rangle = +1$, with $|0\rangle$ denoting the magnon vacuum. This indicates that a quantum of α-magnon carries -1 spin angular momentum, and a quantum of β-magnon carries a $+1$ spin angular momentum along the $\hat{z}$ direction [7].

Unlike the ferromagnet case, where the ground state energy was the classic result $-NJS^2z/2$, the Hamiltonian H_1 contains a quantum zero-point energy.

$$\Delta E = E_0 - E_{\text{class}} = \sum_k \omega_k - NJSz/2 = JSz\sum_k\left(\sqrt{1 - \gamma_k^2} - 1\right). \tag{2.46}$$

Note that ΔE is negative, i. e., quantum fluctuations reduce the energy of the antiferromagnet. In the ground state of H_1 there are no magnons. The ground state is the vacuum of the α and β bosons. However, the ground state admits configurations with arbitrary numbers of spin flips relative to the Néel state, which reduces the long-range Néel order at $T = 0$.

In the antiferromagnet, the order parameter is the staggered magnetization, defined by the average of $\langle \vec{S}_i \rangle$ over the sites of one of the two sublattices. For sublattice A we have

$$m_A = \frac{1}{N_A} \sum_{i \in A} \langle S_i^z \rangle = S - \frac{1}{N_A} \sum_{i \in A} \langle a_i^\dagger a_i \rangle = S - \frac{1}{N_A} \sum_k \langle a_k^\dagger a_k \rangle. \tag{2.47}$$

The correction to the classical result S is therefore given by

$$\Delta m_A = \frac{1}{N_A} \sum_k \langle a_k^\dagger a_k \rangle. \tag{2.48}$$

Using (2.37) we find

$$\Delta m_A = \frac{2}{N} \sum_k \left[u_k^2 \langle \alpha_k^\dagger \alpha_k \rangle + v_k^2 \left\langle \beta_{-k} \beta_{-k}^\dagger \right\rangle + u_k v_k \left(\left\langle \alpha_k \beta_{-k} + \alpha_k^\dagger \beta_{-k}^\dagger \right\rangle \right) \right]. \tag{2.49}$$

We have the relation

$$\langle \alpha_k^\dagger \alpha_k \rangle = \left\langle \beta_k^\dagger \beta_k \right\rangle = n_k, \tag{2.50}$$

where n_k is the Bose factor. For the other terms, let us consider

$$\langle \alpha_k \beta_{-k} \rangle = \frac{1}{Z} \sum_n e^{-\beta E_n} \langle n | \alpha_k \beta_{-k} | n \rangle. \tag{2.51}$$

The term $\langle n | \alpha_k \beta_{-k} | n \rangle$ is an overlap of the state $\beta_{-k} | n \rangle$ and $\alpha_k^\dagger | n \rangle$. Both states are eigenstates of H_1, but as they do not have identical sets of occupation numbers of the various α and β bosons, their overlap is zero. Hence $\langle \alpha_k \beta_{-k} \rangle$ and its complex conjugate are zero. Accordingly, we can write

$$\Delta m_A = \frac{2}{N} \sum_k [(u_k^2 + v_k^2) n_k + v_k^2] = -\frac{1}{2} + \frac{2}{N} \sum_k (n_k + 1/2) \left(1 - \gamma_k^2\right)^{-1/2}. \tag{2.52}$$

At zero temperature $n_k = 0$, but we have a correction to m_A even at zero temperature. The integral $\int d^d k (1 - \gamma_k^2)^{-1/2}$ converges for $d = 2$ and 3. We find $\Delta m_A \approx 0.197$ in 2d and $\Delta m_A \approx 0.078$ in 3d [1]. So the system is ordered at $T = 0$. Note that $\Delta m_A = \langle n_i \rangle$, and we see that $\langle n_i \rangle \ll 2S$ even for $S = ½$. So the use of the linear spin-wave theory is justified. However, the integral diverges in 1d. That is, the one-dimensional antiferromagnet (AF) is disordered even at $T = 0$. The one-dimensional antiferromagnet's ground state is very complicated and is different for S integer and half-odd integers. I will not consider this case here. For details, see [8, 9].

Some authors introduce the Fourier transform of the boson operators with different signs in the two sublattices' exponentials. That is, they use

$$a_i = \frac{1}{\sqrt{N_A}} \sum_k e^{-i\vec{k}\cdot\vec{r}_i} a_k, \quad b_j = \frac{1}{\sqrt{N_B}} \sum_k e^{+i\vec{k}\cdot\vec{r}_j} b_k. \tag{2.53}$$

The quadratic part of H is then given by

$$H_1 = JSz \sum_k [\gamma_k (a_k b_k + b_k^\dagger b_k) + (a_k a_k^\dagger + b_k^\dagger b_k)], \tag{2.54}$$

and the Bogoliubov transformation is now

$$a_k = u_k \alpha_k + v_k \beta_k^\dagger, \quad b_k = u_k \beta_k + v_k \alpha_k^\dagger. \tag{2.55}$$

However, all physical quantities will be the same as before.

2.3 Helimagnets

In this section, following Blundell [10], I will consider the case of a linear chain along the z-direction with a nearest-neighbor exchange constant J_1 and a next nearest-neighbor exchange constant J_2. We can imagine that we also have ferromagnetic couplings in the x and y-directions, in which case we are treating layers in a 3d model. Denoting by θ, the angle between nearest-neighbor spins in the z-direction, the energy of the coupling along the chain in the classical limit will be given by

$$E = -2NS^2(J_1 \cos\theta + J_2 \cos 2\theta), \tag{2.56}$$

where N is the number of atoms in each xy-plane. Minimizing the energy $\partial E/\partial\theta = 0$, we get

$$(J_1 + 4J_2 \cos\theta)\sin\theta = 0. \tag{2.57}$$

The solutions are $\sin\theta = 0$, which implies $\theta = 0$, or $\theta = \pi$ (ferro or antiferro) or

$$\cos\theta = -J_1/4J_2. \tag{2.58}$$

This last solution corresponds to the helical order (helimagnetism). It is favored over either ferromagnetism or antiferromagnetism when $J_2 < 0$ and $|J_1| < 4|J_2|$. The pitch of the spiral along the z-direction will not, in general, be commensurate with the lattice spacing.

2.4 Rotated sublattice

Some authors (for instance [11]), when studying antiferromagnets, rotate the axis in spin space for sites on the down sublattice, so that the local z-axis is aligned with the spin direction in the classical Néel state. In this case, we define the rotated spins $\tilde{S}$ in the sublattice B as

$$\tilde{S}_j^z = -S_j^z, \quad \tilde{S}_j^x = S_j^x, \quad \tilde{S}_j^y = -S_j^y, \quad j \in B. \tag{2.59}$$

Note that this transformation preserves the commutation relations. After the transformation, the Hamiltonian becomes

$$H = -J\sum_{\langle ij\rangle} S_i^z \tilde{S}_j^z + \frac{J}{2}\sum_{\langle ij\rangle}\left(S_i^+ \tilde{S}_j^+ + S_i^- \tilde{S}_j^-\right). \tag{2.60}$$

We use the same HP representation as given by (2.27) for S on sublattice A, and $\tilde{S}$ on sublattice B. In the staggered frame of reference, the classical ground state is now ferromagnetic, and we do not need to introduce two species of bosons. We find

$$H_1 = JSz\sum_k\left[a_k^\dagger a_k + \frac{\gamma_k}{2}(a_k^\dagger a_{-k}^\dagger + a_k a_{-k})\right], \tag{2.61}$$

and the Bogoliubov transformation is

$$a_k = \cosh\theta_k \alpha_k + \sinh\theta_k \alpha_{-k}^\dagger. \tag{2.62}$$

Following the same procedure as we did before, we obtain for the quadratic term

$$H_1 = \sum_k \omega_k(\alpha_k^\dagger \alpha_k + 1/2), \tag{2.63}$$

with ω_k given by (2.43). The rotated sublattice formalism can be used in several other lattices, as I discuss in the next chapter. When we use two boson-types to treat the antiferromagnet, we get two bands, each living in the reduced magnetic Brillouin zone, with the bands labeled by $S_{tot}^z = \pm 1$. By symmetry, the two bands coincide. In the rotated sublattice approach, we get only one band, but this lives on the original Brillouin zone (which, as we saw, is twice as large as the reduced zone), so the state counting is consistent. However, we can no longer associate a well-defined ± 1 to the magnons.

Let us calculate the ground state of the antiferromagnet on a square lattice. For simplicity, I use the rotated sublattice approach. The ground state is the vacuum $|\psi_0\rangle$ of the α bosons, that is

$$\alpha_k|\psi_0\rangle = 0. \tag{2.64}$$

It is easier to consider a simple problem with an operator b, such that $[b, b^\dagger] = 1$. We define the operator

$$S = \frac{x}{2}[b^2 - (b^\dagger)^2]. \tag{2.65}$$

We have

$$[S, b] = xb^\dagger \text{ and } [S, b^\dagger] = xb. \tag{2.66}$$

Now, we calculate

$$\gamma = e^S b e^{-S}. \tag{2.67}$$

Expanding equation (2.67) in power series, we get

$$e^S b e^{-S} = b + [S, b] + \frac{1}{2}[S, [S, b]] + \ldots = b + xb^\dagger + \frac{1}{2}x^2 b + \ldots \tag{2.68}$$

Calculating higher-order terms, we obtain

$$\gamma = \cosh(x)b + \sinh(x)b^\dagger. \tag{2.69}$$

We can follow the same procedure for boson operators to show that the Bogoliubov transformation (2.62) can be written as

$$\alpha_k = U a_k U^{-1}, \tag{2.70}$$

where

$$U = \exp\left[\sum_{k\neq 0} \frac{\theta_k}{2}(a_k^\dagger a_k^\dagger - a_k a_k)\right]. \tag{2.71}$$

Denoting the vacuum of the a_k bosons by $|0\rangle$, we have $a_k|0\rangle = 0$. Hence we can write

$$U^{-1}\alpha_k U|0\rangle = 0 \to |\psi_0\rangle = U|0\rangle, \tag{2.72}$$

which can be written as

$$|\psi_0\rangle = c \exp\left[\frac{1}{2}\sum_{k\neq 0} \theta_k a_k^\dagger a_k^\dagger\right]|0\rangle, \tag{2.73}$$

where c is a renormalization constant.

2.5 The XY model

In this section, I will treat the two-dimensional Hamiltonian

$$H = -J\sum_{\langle i,j\rangle}\left(S_i^x S_j^x + S_i^y S_j^y + \lambda S_i^z S_j^z\right), \tag{2.74}$$

with $0 \leqslant \lambda < 1$. For $\lambda = 0$ we have the XY model. Note that there is a difference between the classical XY and the quantum XY model, although the same Hamiltonian describes both models. However, in the classical model (sometimes called the rotor model), the spins have only two components, $\vec{S} = S(\cos\phi,\ \sin\phi)$, whereas in the quantum case, they have all three components $\vec{S} = (S^x, S^y, S^z)$ and $\langle (S^z)^2\rangle \neq 0$. The spins in Hamiltonian (2.74) minimize the energy when they are in the XY plane, and therefore, the physics will be the same as the pure XY model. In chapter 6, we will study Hamiltonian (2.74) with a single site anisotropy.

A convenient representation to study Hamiltonian (2.74) was introduced by Villain [12] as

$$S_r^+ = e^{i\phi_r}\sqrt{(S + 1/2)^2 - (S_r^z + 1/2)^2},\ S_r^- = \sqrt{(S + 1/2)^2 - (S_r^z + 1/2)^2}\, e^{-i\phi_r}, \tag{2.75}$$

where ϕ is the operator corresponding to the azimuthal angle of the spin around the z-axis. We treat the antiferromagnet in the same way, with the substitution $\phi_r \to \phi_r + \pi$ in alternate sites. For finite S, ϕ_r is a Hermitian operator defined by

$$\left[\phi_r, S_{r'}^z\right] = i\delta_{rr'}, \quad [\phi_r, \phi_{r'}] = 0. \tag{2.76}$$

From (2.76) it easy to check that the operators defined by (2.75) verify the relations:

$$[S_r^z, S_{r'}^{\pm}] = \pm\delta_{rr'}S_r^z, \quad [S_r^+, S_{r'}^-] = 2\delta_{rr'}S_r^z. \tag{2.77}$$

Taking (2.75) into (2.74) we get

$$H = -\frac{J}{2}\sum_{r,a}\left[\sqrt{(S + 1/2)^2 - (S_r^z + 1/2)^2}\cos(\phi_r - \phi_{r+a}) \right. \\ \left. \sqrt{(S + 1/2)^2 - (S_{r+a}^z + 1/2)^2} + \lambda S_r^z S_{r+a}^z\right]. \tag{2.78}$$

Expanding equation (2.78) in terms of $(S_r^z)^2$ and $(\phi_r - \phi_{r+a})$ we obtain

$$H_1 = \frac{J}{2}\sum_{r,a}\left[\tilde{S}^2\left(\phi_r^2 - \phi_r\phi_{r+a}\right) + (S_z^z)^2 - \lambda S_r^z S_{r+a}^z\right], \tag{2.79}$$

where $\tilde{S}^2 = S(S + 1)$. Fourier transforming, we get

$$H_1 = 2J\sum_q\left[\tilde{S}^2(1 - \gamma_q)\phi_q\phi_{-q} + (1 - \lambda\gamma_q)S_q^z S_q^z\right], \tag{2.80}$$

where $\gamma_q = \frac{1}{2}(\cos q_x + \cos q_y)$. We can write (2.80) as

$$H_1 = 2\sum_q\left[\tilde{S}^2 a(q)\phi_q\phi_{-q} + b(q)S_q^z S_{-q}^z\right], \tag{2.81}$$

with

$$a(q) = J(1 - \gamma_q), \; b(q) = J(1 - \lambda\gamma_q). \tag{2.82}$$

By introducing the canonical transformation

$$\phi_q = \left(\frac{b(q)}{a(q)}\right)^{1/4}(a_q^\dagger + a_{-q}), \quad S_q^z = i\left(\frac{a(q)}{b(q)}\right)^{1/4}(a_q^\dagger - a_{-q}), \tag{2.83}$$

where $a_q^\dagger$ and a_q are the boson-creation and annihilation operators, respectively, we can write equation (2.81) as (note that here $a(q)$ and a_q are different objects)

$$H_1 = \sum_q \omega_q(a_q^\dagger a_q + 1/2), \tag{2.84}$$

where

$$\omega_q = 2\tilde{S}\sqrt{a(q)b(q)}\,. \tag{2.85}$$

Using equation (2.83), static correlations can be calculated, and the result is:

$$\langle (S_r^z)^2 \rangle = \frac{\tilde{S}}{2\pi^2}\int_0^{\pi}\int_0^{\pi} d^2q \sqrt{\frac{a(q)}{b(q)}}\coth\left(\frac{\beta\omega_q}{2}\right), \tag{2.86}$$

$$\langle \phi_q \phi_{-q} \rangle = \frac{1}{2\tilde{S}}\sqrt{\frac{b(q)}{a(q)}}\coth\left(\frac{\beta\omega_q}{2}\right). \tag{2.87}$$

Now I will calculate the correlation function given by

$$g(r) = \langle \exp\{i[\phi(\vec{r}) - \phi(0)]\} \rangle. \tag{2.88}$$

To do the calculation, I will take the low k limit so that

$$\langle \phi_k \phi_{-k} \rangle \approx \frac{k_B T}{JS^2 k^2 a^2}, \tag{2.89}$$

where I have put explicitly the lattice constant a. Equation (2.88) can then be written as

$$g(r) = \exp\left[-\frac{k_B T}{\pi^2 JS^2}\int_0^{\pi/a}\int_0^{\pi/a}\left(\frac{1 - \cos\vec{k}\cdot\vec{r}}{k^2}\right)d^2\vec{k}\right]. \tag{2.90}$$

Carrying out the integration in polar coordinates and using

$$\int_0^{2\pi} d\theta\, \cos(kr\cos\theta) = 2\pi J_0(kr), \tag{2.91}$$

where J_0 is the zeroth-order Bessel function, we obtain

$$g(r) = \exp\left[-\frac{k_B T}{2\pi JS^2}\int_0^{\pi/a}\frac{1 - J_0(kr)}{k}dk\right]. \tag{2.92}$$

Writing $x = kr$, the integral becomes

$$\int_0^{\pi r/a}\frac{1 - J_0(x)}{x}dx. \tag{2.93}$$

For large r/a, the dominant contribution comes from the region $x \gg 1$, in which case we can ignore the Bessel function. The final result is

$$g(r) \approx \exp\left[-\frac{k_B T}{2\pi JS^2}\ln\left(\frac{\pi r}{a}\right)\right] = \left(\frac{\pi r}{a}\right)^{-\eta}, \tag{2.94}$$

where

$$\eta = \frac{k_B T}{2\pi J S^2}. \tag{2.95}$$

The correlation function falls off algebraically at all finite temperatures, and there is no long-range order in the system. In the above calculations, we have relaxed the constraint that ϕ must be in the range 0 to 2π and allowed it to range from $-\infty$ to ∞. Now we consider ϕ an angle variable. In the following, I will consider only slowly varying configurations, that is, those with adjacent angles nearly equal. At low temperatures, thermal fluctuations are not so large, and the spin direction change is gradual. Expanding the Hamiltonian up to quadratic terms, taking $\lambda = 0$, $a = 1$ and using

$$\frac{\partial \phi_n}{\partial t} = -i[\phi_n, H] = -i2JS_n^z, \tag{2.96}$$

we get in the continuum limit

$$H = \int dxdy \left\{ \frac{1}{2J}\left(\frac{\partial \phi}{\partial t}\right)^2 + J\tilde{S}^2\left[\left(\frac{\partial \phi}{\partial x}\right)^2 + \left(\frac{\partial \phi}{\partial y}\right)^2\right]\right\}, \tag{2.97}$$

neglecting a constant term.

The equation of motion is

$$\frac{\partial^2 \phi}{\partial^2 x} + \frac{\partial^2 \phi}{\partial^2 y} - \frac{1}{c^2}\frac{\partial^2 \phi}{\partial t^2} = 0, \tag{2.98}$$

where $c = \sqrt{2}J\tilde{S}$. A static vortex solution is given by

$$\phi = \tan^{-1}(y/x). \tag{2.99}$$

If we consider a vortex configuration, as we go around a closed path containing the center of the vortex, ϕ will change by 2π for each revolution of the spin. The vorticity q is defined as

$$\frac{1}{2\pi}\oint d\phi(r) = q, \quad \text{where} \quad q = 0, \pm 1, \pm 2, \ldots \tag{2.100}$$

For an isolated vortex, the energy is

$$E \approx 2\pi J \ln(R/a), \tag{2.101}$$

where R is the radius of the system, and a is the lattice spacing. The energy increases logarithmically with the system size. The thermal generation of vortices produces an even number of vortices and antivortices. Bound pairs have lower energies than free vortices but have low entropy. The system to minimize the free energy undergoes a topological transition at a critical temperature T_{BKT}, called the Berezinskii–Kosterlitz–Thouless temperature resulting from the unbinding of vortex–antivortex pairs [13, 14]. Note that there is no local order parameter below T_{BKT}.

No spin-wave theory gives T_{BKT}. I will present a simple theory called *self-consistent harmonic approximation* (SCHA) that is adequate for the calculation of T_{BKT} [15].

In the SCHA we expand the square root and the exponential term $e^{i(\phi_i-\phi_j)}$ and use Wick's theorem to write n-operator expressions in terms of two operator expressions, for instance

$$\phi^4 = :\phi^4: +6\langle\phi^2\rangle - 2\langle\phi^2\rangle^2, \tag{2.102}$$

where $:\phi^s:$ denotes normal-ordered operators [16]. The general formula is

$$\phi^{2n} = \sum_{l=0}^{n-l}\langle\phi^2\rangle^l :\phi^{2n-2l}: (2l-1)!!\binom{2n}{2l} - (n-1)(2n-1)!!\langle\phi^2\rangle^n. \tag{2.103}$$

The SCHA drops the normal-ordered interaction term and computes $\langle OO\rangle$ (where $O = \phi, S^z$) self-consistently using the quadratic Hamiltonian.

In the SCHA, the Hamiltonian is written as

$$H_1 = \frac{J}{2}\sum_{r,a}\left[\rho\tilde{S}^2(\phi_r^2 - \phi_r\phi_{r+a}) + (S_r^z)^2 - \lambda S_r^z S_{r+a}^z\right], \tag{2.104}$$

where the stiffness ρ is given by

$$\rho = \left\langle\left[1 - \left(\frac{S_r^z}{\tilde{S}}\right)^2\right]\right\rangle\langle\cos(\phi_r - \phi_{r+a})\rangle. \tag{2.105}$$

Using

$$\langle\cos(\phi_r - \phi_{r'})\rangle = \exp[-\frac{1}{2}\langle(\phi_r - \phi_{r'})^2\rangle], \tag{2.106}$$

we can write

$$\rho = \langle[1 - r]\rangle\exp\left[-\frac{1}{2}\langle(\phi_r - \phi_{r+a})^2\rangle\right], \tag{2.107}$$

where

$$r = \left(\frac{S_r^z}{\tilde{S}}\right)^2. \tag{2.108}$$

Finally, we have

$$\rho = \langle(1 - r)\rangle\exp\left[-\frac{1}{\pi^2}\int_0^\pi\int_0^\pi(1 - \gamma_q)\langle\phi_q\phi_{-q}\rangle\right]. \tag{2.109}$$

We can use the former expressions (2.84), (2.86), (2.87) with the parameter $a(q)$ written as

$$a(q) = 2J\rho(1 - \gamma_q). \tag{2.110}$$

We calculate the stiffness self-consistently as a function of temperature. T_{BKT} is estimated as the temperature where ρ drops discontinuously to zero. In the classical limit, the theory gives good results compared with Monte Carlo simulations [17].

The phase transition at T_{BKT} changes the long-range algebraic order to short-range order. We would like to point out that the BKT transition does not alter any symmetry. Kosterlitz and Thouless studied the classical XY model using renormalization group theory and found that the transition temperature is given by

$$\frac{\pi J}{k_B T_{BKT}} - 2 = \exp\left(-\frac{\pi^2 J}{2k_B T_{BKT}}\right). \tag{2.111}$$

The susceptibility diverges below T_{BKT} and is finite above T_{BKT}. The correlation length behaves as

$$\xi \sim \exp(bt^{-1/2}), \quad t > 0, \tag{2.112}$$

where $b = (T - T_{BKT})/T_{BKT}$, and $b \approx 1.5$.

The XY model has also been used as a model for spin glasses, superconductivity and liquid helium.

2.6 The compass model

The compass model study is an active field of research with still many interesting and open problems. In a more general sense, a compass model can be described as follows [18]: At each site of a lattice, we prescribe degrees of freedom designated by $\vec{\tau}$ that could be spin, pseudospin, color, flavor, or orbital. The interaction between the degrees of freedom has the following properties: (a) there is only one interaction between specific vector components of $\vec{\tau}$, (b) on different bonds in the lattice, different vector components interact. For instance, if a site i is linked to nearest-neighbor sites j and k, the interaction along the lattice link $\langle ij\rangle$ could be of the type $\tau_i^x\tau_j^x$, whereas on the link $\langle ik\rangle$ it is $\tau_i^y\tau_k^y$. The three-dimensional 90° compass model is defined on a cubic lattice as

$$H^{90^\circ} = -J\sum_i(\tau_i^x\tau_{i+\hat{e}_x}^x + \tau_i^y\tau_{i+\hat{e}_y}^y + \tau_i^z\tau_{i+\hat{e}_z}^z). \tag{2.113}$$

The ground state of Hamiltonian (2.113) is non-trivial. Pairs of spins parallel to the x-axis favor $\vec{\tau}$ pointing along x so that $\langle\tau^x\rangle \neq 0$. The same holds for the other directions. At a site, the bonds along x, y, and z cannot be satisfied simultaneously, and the interactions are strongly frustrated. The much-studied Kitaev's honeycomb model is exactly solvable. The lattices links on a honeycomb lattice may point along three different directions designed by $\hat{e}_1$, $\hat{e}_2$ and $\hat{e}_3$, where the angle between the three unit vectors is 120°. The Hamiltonian describing the Kitaev model is given by

$$H_{Kitaev} = -J_x\sum_{e_1\text{bonds}}\tau_i^x\tau_j^x - J_y\sum_{e_2\text{bonds}}\tau_i^y\tau_j^y - J_z\sum_{e_3\text{bonds}}\tau_i^z\tau_j^z. \tag{2.114}$$

In magnetic systems, the compass model can be used to study strongly correlated models with orbital degeneracy, as materials with large spin–orbit coupling may give rise to compass interactions for some lattice structures. For more details about the model, see [18, 19]. Here I will consider the two-dimensional Heisenberg-compass model on a square lattice described by the following Hamiltonian

$$H = \sum_{i,j}[J_x S^x_{i,j} S^x_{i,j+1} + J_z S^z_{i,j} S^z_{i+1,j} + J(S_{i,j} \cdot S_{i,j+1} + S_{i,j} \cdot S_{i+1,j})], \tag{2.115}$$

where J_x couples the S^x spin component along the x-direction and J_z the S^z component along the z-direction, thus the spin components involved in the interactions depend on the bond orientation. J is the exchange interaction between the nearest neighbors. The ground state is highly degenerated when $J = 0$, but this degeneracy is removed by the Heisenberg term, even with an arbitrarily small amplitude, leading to various magnetically ordered phases. This model is physically more simple to be studied than the pure compass model.

When the anisotropic interactions are varied through the isotropic point of the two-dimensional compass model, a quantum phase transition between two different directional order types occurs. At the transition, a discrete symmetry in spin space is spontaneously broken since the frustrated interactions along two different directions are equivalent, and the spin orientation follows one of them [19]. If the compass coupling interactions $|J_x|$ and $|J_z|$ are not equal, and the Heisenberg coupling strength $|J|$ is finite, the ground state is characterized by long-range spin order with an easy axis. In the case of $J = 0$, $J_x = J_z$, there is no conventional ordering. Still, there is, nevertheless, long-ranged order in the form of directional ordering fluctuations that belong to the Ising universality.

Trousselet *et al* [19] have shown that the Hamiltonian (1), depending on the parameters J, J_x, and J_z, has the following phases: F_z, F_x, F_y, C_z, C_z', G_z, G_x, G_y, C_x, C_x'. The index $\alpha = x, y, z$ denotes the easy axis or spin direction favored, while the capital letter indicates the type of order, i.e., G for Néel-type antiferromagnetic phase, F for ferromagnetic phase, and C for columnar phase, i.e., with nearest-neighbor spin correlations being AF for one direction and FM for the other. A prime for the C-type phases indicates that the interactions are AF on bonds where compass interactions couple spin components along the easy axis.

Here I will study the phases G_z, F_z and C_z, using the Holstein–Primakoff representation. The expressions for the other phases can be obtained from these phases using arguments of symmetry.

For a set of parameters J_x, J_y and J, we calculate which phase has the lowest ground state energy. We start with the case where the spins are aligned antiferromagnetically in the z-direction (we have here: $J > 0$, $J_z > 0$, $-(1 + 2J/J_z) \leqslant J_x/J_z \leqslant 1$), that is, we choose the Néel state to be in the $\hat{z}$ and -$\hat{z}$ directions on sublattice A and B, respectively. This phase is the G_z phase. In sublattice B we define the rotated spins as:

$$\tilde{S}^z_n = -S^z_n, \quad \tilde{S}^x_n = S^x_n, \quad \tilde{S}^y_n = -S^y_n, \tag{2.116}$$

where $n = i, j+1$ and $i+1, j$. Using the Holstein–Primakoff representation

$$S^{+}_{i,j} = \sqrt{2S}a_{i,j}, \quad S^{-}_{i,j} = a^{\dagger}_{i,j}\sqrt{2S}, \quad S^{z}_{i,j} = S - a^{\dagger}_{i,j}a_{i,j}, \tag{2.117}$$

in the Hamiltonian (2.115), we obtain

$$\begin{aligned} H_1 = S\sum_k \Big\{ &\frac{J_x}{2}\cos k_x(a_k a_{-k} + a_k a^{\dagger}_k + a^{\dagger}_k a_k + a^{\dagger}_k a^{\dagger}_{-k}) + 2J_z a^{\dagger}_k a_k \\ &+ 4J\Big[a^{\dagger}_k a_k + \frac{\gamma_k}{2}(a^{\dagger}_k a^{\dagger}_{-k} + a_k a_{-k})\Big]\Big\}, \end{aligned} \tag{2.118}$$

where here

$$\gamma_k = \cos k_x + \cos k_z. \tag{2.119}$$

H_1 includes normal and anomalous terms $a_k a_{-k}$ and $a^{\dagger}_k a^{\dagger}_{-k}$. Using the Bogoliubov transformation

$$a_k = \cosh\theta_k \alpha_k + \sinh\theta_k \alpha^{\dagger}_{-k}, \quad a^{\dagger}_k = \cosh\theta_k \alpha^{\dagger}_k + \sinh\theta_k \alpha_{-k}, \tag{2.120}$$

we get

$$\begin{aligned} H = \sum_k \Big[&(b_k \cosh 2\theta_k + c_k \sinh 2\theta_k)\alpha^{\dagger}_k \alpha_k \\ &+ \frac{1}{2}(b_k \sinh 2\theta_k + c_k \cosh 2\theta_k)(\alpha^{\dagger}_k \alpha^{\dagger}_{-k} + \alpha_k \alpha_{-k})\Big], \end{aligned} \tag{2.121}$$

where

$$b_k = S(J_x cos\, k_x + 2J_z + 4J), \quad c_k = S(J_x \cos k_x + 2J\gamma_k). \tag{2.122}$$

To eliminate the term $(\alpha^{\dagger}_k \alpha^{\dagger}_{-k} + \alpha_k \alpha_{-k})$ we choose

$$\tanh 2\theta_k = -c_k/b_k, \tag{2.123}$$

which yields

$$H = \sum_k \omega_k \alpha^{\dagger}_k \alpha_k + \text{Constant}, \tag{2.124}$$

with

$$\omega_k = 2\sqrt{b_k^2 - c_k^2}. \tag{2.125}$$

Using (2.122) we get

$$\omega_k = 2S\sqrt{(J_x \cos k_x + 2J_z + 4J)^2 - (J_x \cos k_x + 2J\gamma_k)^2}. \tag{2.126}$$

Now we consider the case where $J < 0$, $J_z < 0$ and $-1 + 2J/J_z < J_x/J_z < 1$. This phase is the F_x phase. Following the same procedure as we did above, and remembering that now we do not need to rotate the spins in the B sublattice, we obtain:

$$b_k = S[-(2J_z+4J) + 2J\gamma_q + J_x \cos k_x], \quad c_k = SJ_x \cos k_x. \tag{2.127}$$

The spin-wave Hamiltonian contains $a_i a_j$ type terms besides $a_i^\dagger a_j$ type, which leads to a break of the SU(2) invariance. As a result, unlike the ferromagnetic Heisenberg model, the spin-wave dispersion has a square-root form as in the antiferromagnetic Heisenberg case. For J_z, $J_x \to 0$, one recovers the result for the ferromagnet, as it should be. At $J_z = J_x$, the gap vanishes at $\boldsymbol{q} = (0,0)$, signaling a transition to the F_x phase.

As a final example, we consider the C_z phase that takes place for $J > 0$, $J_z < 0$, and $-1 + 2J/J_z < J_x/J_z < 1$. Now we have:

$$b_k = S(2|J_z| + J_x \cos k_x + 2J \cos k_z), \quad c_k = S(2J + J_x)\cos k_x. \tag{2.128}$$

2.7 The Jordan–Wigner transformation

The Jordan–Wigner representation is mainly used to treat spin ½ one-dimensional chains in terms of fermionic operators. Using fermions, we do not need any constraint to prevent two particles from occupying the same site: the Pauli exclusion principle ensures it. The generalization of the Jordan–Wigner for a higher dimension is presented in ref. [9]. In dimensions higher than 1, it transforms local interactions into non-local ones. The Jordan–Wigner transformation is also used in the study of the Kitaev model discussed in the last section. The representation is given by

$$S_i^+ = c_i^\dagger \exp\left(i\pi\sum_{j=1}^{i-1} c_j^\dagger c_j\right), \quad S_i^- = \exp\left(-i\pi\sum_{j=1}^{i-1} c_j^\dagger c_j\right) c_i, \quad S_i^z = c_i^\dagger c_i - 1/2, \tag{2.129}$$

where c_i is a fermionic operator satisfying anticommutation relations:

$$\{c_i, c_j^\dagger\} = c_i c_j^\dagger + c_j^\dagger c_i = \delta_{ij}, \qquad \{c_i, c_j\} = \{c_i^\dagger, c_j^\dagger\} = 0. \tag{2.130}$$

The factor $\exp(\pm i\pi\sum_{j=1}^{i-1} c_j^\dagger c_j)$ introduces a phase, which depends on the number of particles to the left of a given state. Taking (2.129) into the ferromagnetic Heisenberg Hamiltonian, we get

$$\begin{aligned} H &= -J\sum_i\left(\frac{S_i^+ S_{i+1}^- + S_i^- S_{i+1}^+}{2} + \lambda S_i^z S_{i+1}^z\right) \\ &= -J\sum_i\left[\frac{1}{2}c_i^\dagger e^{-i\pi c_i^\dagger c_i} c_{i+1} + \frac{1}{2}c_i c_{i+1}^\dagger e^{i\pi c_i^\dagger c_i} + \lambda\left(c_i^\dagger c_i - \frac{1}{2}\right)\left(c_{i+1}^\dagger c_{i+1} - \frac{1}{2}\right)\right] \\ &= -J\sum_i\left[\frac{1}{2}c_i^\dagger c_{i+1} + \frac{1}{2}c_i c_{i+1}^\dagger(-1) + \lambda\left(c_i^\dagger c_i - \frac{1}{2}\right)\left(c_{i+1}^\dagger c_{i+1} - \frac{1}{2}\right)\right] \\ &= -\frac{J}{2}\sum_i\left\{c_i^\dagger c_{i+1} + c_{i+1}^\dagger c_i + 2\lambda\left(c_i^\dagger c_i - \frac{1}{2}\right)\left(c_{i+1}^\dagger c_{i+1} - \frac{1}{2}\right)\right\}. \end{aligned} \tag{2.131}$$

For the XY model ($\lambda = 0$), we get a free fermions system with an exact solution. Writing

$$c_j = \frac{1}{\sqrt{N}}\sum_k e^{ikj}c_k, \tag{2.132}$$

we get

$$H_{XY} = -\frac{J}{N}\sum_k \varepsilon_k c_k^\dagger c_k, \tag{2.133}$$

with $\varepsilon_k = J\cos k$, and $k \in [-\pi, \pi]$. In the ground state, fermion states with $\varepsilon_k < 0$ are occupied (that is, states with $|k| \leqslant |k_F| = \pi/2$), and those with $\varepsilon_k > 0$ are empty. The simplest such excitation involves creating a particle-hole pair on top of the ground state. We see then that the excitations are different from the spin waves treated before. If we measure the energy of the elementary excitations in relation to $k_F = \pm\pi/2$, we have

$$\omega_k = |-J\cos(k \pm \pi/2)| = |J\sin k|. \tag{2.134}$$

The XY model is one of the rare examples where we can calculate the exact results. For instance, we find [20]

$$\langle S_i^z S_{i+n}^z\rangle = \frac{1}{2\pi^2 n^2}[\cos(\pi n) - 1]. \tag{2.135}$$

We see that the S^z correlation function decays algebraically. The evaluation is straightforward because the operator involved is a density operator in the fermion representation. For the other components S^x and S^y, the results have a much more complicated structure.

2.8 Hardcore bosons

As was said before, the linear spin-wave theory demands $\langle n_i\rangle \ll 2S$, and therefore is best suited for large values of spins, for magnetically ordered phases and low temperatures. In the case of a spin ½ model, one can use the Matsubara–Matsuda transformation into hardcore bosons, where the spin operators are expressed in terms of bosonic operators with at most one boson per site:

$$S_i^+ = b_i^\dagger,\quad S_i^- = b_i,\quad S_i^z = b_i^\dagger b_i - 1/2 = n_i - 1/2. \tag{2.136}$$

The spin ½ operators obey (besides equation 1.2) an on-site exclusion principle

$$S_i^+S_i^- + S_i^-S_i^+ = 1,\quad (S_i^+)^2 = (S_i^-)^2 = 0. \tag{2.137}$$

The hardcore operators should then satisfy the commutation relation

$$[b_i, b_j^\dagger] = \delta_{ij}(1 - 2b_i^\dagger b_j). \tag{2.138}$$

The hardcore constraint can be realized by including in the Hamiltonian an infinite on-site repulsion

$$H_\infty = \frac{U}{2}\sum_i b_i^\dagger b_i^\dagger b_i b_i, \quad \text{with } U \to \infty. \tag{2.139}$$

Bosonic distribution allows us to have more than a particle per site. Here it is possible to have two or more bosons in the same state, but the respective required energy is infinite. Thus, we have a group of particles whose creation and annihilation operators satisfy the bosonic commutation rules but respect the fermionic distribution.

We can use the mapping (2.136) for any spin ½ model in any dimension. The disadvantage of the method is that it is difficult to treat analytically. Due to the strong interaction, we have to treat the infinite on-site interaction non-perturbatively. Fauseweh *et al* [21] have shown that this interaction can be treated within a self-consistent ladder approximation when an energy gap separates the excited states from the ground state.

2.9 Majorana fermions

Spin ½ can also be represented by the Hermitian Majorana fermions. These operators all anticommute with each other

$$\left\{\gamma_i^\alpha, \gamma_j^\beta\right\} = 2\delta_{\alpha\beta}\delta_{ij}, \quad \gamma_i^\dagger = \gamma_i, \quad (\gamma_i^\alpha)^2 = 1, \quad \alpha = x, y, z. \tag{2.140}$$

The spin operators are written as

$$S_i^\mu = \frac{i}{4}\varepsilon^{\mu\alpha\beta}\gamma_i^\alpha\gamma_i^\beta. \tag{2.141}$$

It is easy to verify that the spin commutation relations are reproduced, together with the constraint on the spin ½ representation: $\vec{S}^2 = 3/4$. There is a local Z_2 gauge invariance in this representation since changing the sign of $\vec{\gamma}_i$ does not change $\vec{S}_i$. This representation has been used extensively in recent years to treat spin-liquid states on different types of lattices [22].

2.10 Exercises

2.1. A ferrimagnetic chain is an antiferromagnetic chain with alternate spins s and S. We can write the Hamiltonian as

$$H = J\sum_i(\vec{S}_i \cdot \vec{s}_i + \vec{s}_i \cdot \vec{S}_{i+1}).$$

Using the linear spin-wave formalism, calculate the dispersion relation for this chain. There are two types of magnons: a gapless 'acoustical' branch and a gapped 'optical' branch. See [23].

2.2. Consider the quantum Heisenberg model with long-range ferromagnetic interactions decaying as J/r^p in one dimension. The Hamiltonian is then given by

$$H = -J\sum_{i<j}(r_{ij})^{-p}\vec{S}_i \cdot \vec{S}_j.$$

(a) Show that H can be written as

$$H = -\frac{1}{2}\sum_{m=1}^{N}\sum_{n}^{N-1} J\left[\frac{\pi/N}{\sin(\pi n/N)}\right]^p \vec{S}_m \cdot \vec{S}_{m+1}.$$

(b) Calculate the dispersion relation using the Holstein–Primakoff representation.

(c) For what values of p is the magnetization nonnull at zero temperature?

2.3. Show that for the ferromagnetic simple cubic, body-centered cubic, and face-centered cubic with lattice constant a, the Hamiltonian reduces to

$$H = -JNzS^2 + 2JS\sum_k (ka)^2 a_k^\dagger a_k,$$

for small wavelength $|\vec{k} \cdot \vec{\delta}| \ll 1$.

2.4. Prove equation (2.23).

2.5. Consider the two-dimensional antiferromagnetic Heisenberg model with next neighbor interaction J_1 and next near-neighbor interaction (along the diagonals) J_2. For $J_2 \ll J_1$ we can take the Néel state as the ground state. Using the Holstein–Primakoff representation, calculate the dispersion relation for this model. For what value of J_2 does the calculation break down?

Note: This is a frustrated system, as will be discussed in chapter 5. However, it is known that for J_2 below a critical value, the system has Néel order at zero temperatures.

2.6. Prove equation (2.61).

2.7. Prove equations (2.86) and (2.87).

2.8. Calculate the spin-wave dispersion for the Heisenberg ferromagnet in one dimension using a semiclassical approach, as follows:

(a) Calculate the time dependence of $\langle \vec{S}_i \rangle$ using the equation of motion

$$i\hbar\frac{d\langle \vec{S}_i \rangle}{dt} = \langle [\vec{S}_i, H] \rangle.$$

(b) Treat now the spins as classical vectors. The ground state of the system has all the spins aligned, say along the z-direction. Then $S_j^z = S$, $S_j^x = S_j^y = 0$. Consider a state that is a small departure from this state with $S_j^z \approx S$, S_j^x, $S_j^y \ll S$. Obtain the equation of motion.

(c) The normal mode solutions are given by $S_j^x = Ae^{i(q_j a - \omega t)}$, $S_j^y = Be^{i(q_j a - \omega t)}$. Show that $A = iB$. Find then ω.

References

[1] Kittel C 1963 *Quantum Theory of Solids* (New York: Wiley)

[2] Fazekas P 2003 *Lectures Notes on Electron Correlation and Magnetism* (Singapore: World Scientific)

[3] Powalsk M, Schmidt K P and Uhrig G S 2018 Mutually attracting spin waves in the square lattice quantum antiferromagnet *SciPost Phys* **4** 001–38

[4] Stancel D D and Prabhakar A 2009 *Spin Waves Theory and Applications* (Berlin: Springer)

[5] Timm C 2015 *Theory of Magnetism* (Dresden: Technische Universitat Dresden) https://obelix.physik.uni-bielefeld.de/~schnack/molmag/material/timm-lecturenotes.pdf

[6] Manousakis E 1991 The spin ½ Heisenberg antiferromagnet on a square lattice and its application to the cuprous oxides *Rev. Mod. Phys.* **63** 1–62

[7] Cheng R, Okamoto S and Xiao D 2016 Spin Nernst effect of magnons in collinear antiferromagnets *Phys. Rev. Lett.* **117** 217202

[8] Fradkin E 1991 *Field Theories of Condensed Matter Systems* (Redwood City, CA: Addison-Wesley Publishing Company)

[9] Tsvelik A M 1996 *Quantum Field Theory in Condensed Matter* (Cambridge: Cambridge University Press)

[10] Blundell S 2003 *Magnetism in Condensed Matter* (Oxford: Oxford University Press)

[11] Auerbach A 1994 *Interacting Electrons and Quantum Magnetism* (Berlin: Springer)

[12] Villain J 1974 Quantum theory of one- and two-dimensional ferro-and antiferromagnets with an easy magnetization plane *J. Phys.* **35** 27–47

[13] Kosterlitz J M and Thouless D J 1973 Ordering, metastability and phase transitions in two-dimensional systems *J. Phys.: Solid State Phys* **6** 1181–203

[14] Berezinskii V L 1971 Destruction of long-range order in one-dimensional and two-dimensional systems having a continuous symmetry group *Sov. Phys. JETP* **32** 493–500

[15] Pires A S T 2007 Quantum- phase transition in an XY model *Physica* A **373** 387–91

[16] Samuel S 1982 Perturbative analysis of the planar XY model *Phys. Rev.* B **25** 1755–58

[17] Pires A S T 1995 Perturbative analysis of the classical two-dimensional XY model *Solid State Commun.* **96** 915–18

[18] Nussinovand Z and van den Brink J 2015 Compass and Kitaev models: theory and physical motivations *Rev. Mod. Phys.* **87** 1–59

[19] Trousselet F, Oles A M and Horsch P 2010 Compass-Heisenberg model on the square lattice: spin order and excitations *Eur. Lett.* **91** 40005

[20] Niemeijer T 1967 Some exact calculations on a chain of spins ½ *Physica* **36** 377–419

[21] Fauseweh B, Stolze J and Uhrig G S 2014 Finite-temperature line shapes of hardcore bosons in quantum magnets: a diagrammatic approach tested in one dimension *Phys. Rev.* B **90** 024428

[22] Biswas R R, Fu L, Laumann C R and Sachdev S 2011 SU(2)-invariant spin liquids on the triangular lattice with spinful Majorana excitations *Phys. Rev.* B **83** 245131

[23] Pati S K, Ramasesha S and Sen D 1997 Low-lying excited states and low-temperature properties of an alternating spin-1-spin−1/2 chain: a density matrix renormalization group *Phys. Rev.* B **55** 8894–904

Chapter 3

Spin waves II

In this chapter, I use the rotated sublattice technique to study the antiferromagnetic triangular lattice and the antiferromagnetic square lattice in the presence of an external magnetic field. I introduce the Dzyaloshinskii–Moriya interaction, treat the nonlinear spin-wave theory in a mean-field approach, and finally present a modified spin-wave theory.

3.1 Triangular lattice

The spins on a classical Heisenberg antiferromagnetic triangular lattice cannot orient along the direction normal to the lattice plane because of frustration (as will be studied in chapter 5) as three spins on neighboring sites cannot be all antiparallel. Hence, they should lie in the plane of the lattice. The classical model orders in a 120° structure, as shown in figure 3.1 [1, 2]. Calculations using various numerical methods have suggested that the 120° magnetic system remains stable even for S = ½, at $T = 0$. In the triangular lattice, the distance between the two nearest-neighbor spins of the same sublattice is $\sqrt{3}$ of the lattice spacing, which is here taken as unity. We transform into a rotating frame with the z-axis pointing along the local spin direction to use the spin-wave formalism. Then, one needs to define only a single species of boson operators, as was done in section 2, with the wave vectors belonging to the full paramagnetic Brillouin zone (see figure 3.2). The reader can find the Brillouin zone of all kinds of lattices in the literature.

We suppose that the spins lie in the x–z plane so that we can choose the quantization direction along the z-direction just for convenience and transform the Hamiltonian from the laboratory frame (x_0, z_0) to the rotating frame (x, z):

$$S_i^{z_0} = S_i^z \cos\theta_i - S_i^x \sin\theta_i, \quad S_i^{x_0} = S_i^z \sin\theta_i + S_i^x \cos\theta_i, \tag{3.1}$$

with $S_i^{y_0} = S_i^y$. The angle θ is given by $\theta_i = \vec{Q} \cdot \vec{r}_i$, where $\vec{Q} = (4\pi/3, 0)$ is the magnetic ordering wavevector of the 120° spin structure.

doi:10.1088/978-0-7503-3879-0ch3

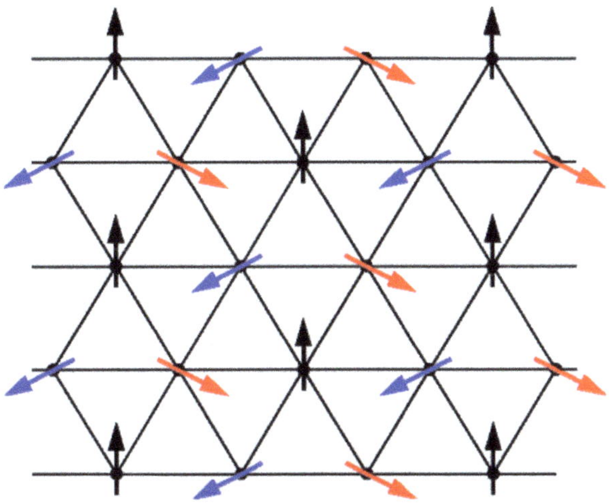

Figure 3.1. The triangular lattice.

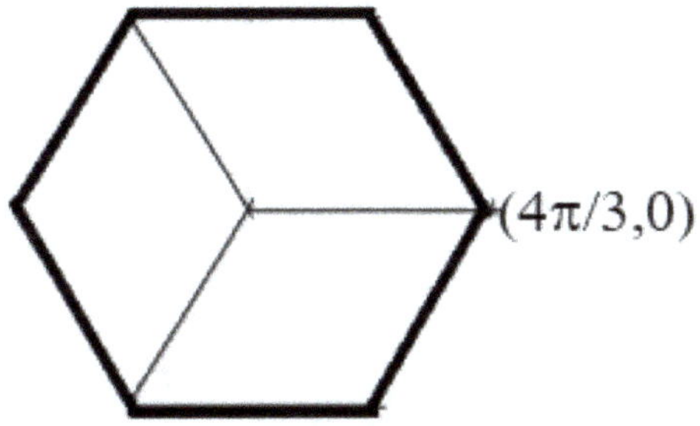

Figure 3.2. The Brillouin zone of the triangular lattice.

The Hamiltonian after this transformation is given by

$$\begin{aligned} H = J\sum_{\langle ij\rangle}\Big[&S_i^yS_j^y + \cos(\theta_i - \theta_j)\big(S_i^zS_j^z + S_i^xS_j^x\big) \\ &+ \sin(\theta_i - \theta_j)\big(S_i^zS_j^x - S_i^xS_j^z\big)\Big]. \end{aligned} \tag{3.2}$$

Now we use the HP representation to first order as we did before. Terms that are linear in a_i and $a_i^\dagger$ vanish because the 120° spin structure corresponds to a minimum of the classical energy. After a Fourier transformation for the bosonic operators, we get the quadratic part of the Hamiltonian

$$H_1 = \sum_k\left[A_k a_k^\dagger a_k - \frac{B_k}{2}\left(a_k^\dagger a_{-k}^\dagger + a_{-k}a_k\right)\right], \tag{3.3}$$

where

$$A_k = 3JS\left(1 + \frac{1}{2}\gamma_k\right), \quad B_k = \frac{9}{2}JS\gamma_k. \tag{3.4}$$

The near-neighbor sites of a spin located at the origin are given by:

$$\vec{\delta}_1 = (1, 0), \quad \vec{\delta}_2 = \frac{1}{2}(1, \sqrt{3}), \quad \vec{\delta}_3 = \frac{1}{2}(-1, \sqrt{3}),$$

$$\vec{\delta}_4 = (-1, 0), \quad \vec{\delta}_5 = \frac{1}{2}(-1, -\sqrt{3}), \quad \vec{\delta}_6 = \frac{1}{2}(1, -\sqrt{3}).$$

Hence

$$\begin{aligned}\gamma_k &= \frac{1}{6}\sum_{\delta} e^{i\vec{k}.\vec{\delta}} = \frac{1}{6}\Big(e^{ik_x} + e^{\frac{i}{2}(k_x+\sqrt{3}k_z)} + e^{\frac{i}{2}(-k_x+\sqrt{3}k_z)} \\ &+ e^{-ik_x} + e^{\frac{i}{2}(-k_x-\sqrt{3}k_z)} + e^{\frac{1}{2}(k_x-\sqrt{3}k_z)}\Big) = \frac{1}{3}\left(\cos k_x + 2\cos\frac{k_x}{2}\cos\frac{\sqrt{3}}{2}k_z\right).\end{aligned} \tag{3.5}$$

Note that some authors use a triangular lattice rotated by 90° and therefore with a different γ_k. Using

$$a_k = u_k\alpha_k + v_k\alpha_{-k}^{\dagger}, \tag{3.6}$$

we obtain

$$H_1 = \sum_k \omega_k\left(\alpha_k^{\dagger}\alpha_k + \frac{1}{2}\right), \tag{3.7}$$

with

$$\omega_k = \sqrt{A_k^2 - B_k^2} = 3JS\sqrt{(1 - \gamma_k)(1 + 2\gamma_k)}. \tag{3.8}$$

This spin-wave spectrum has three zero modes. One at the center of the zone $\vec{k} = (0, 0)$, and the other two at the corners of the BZ, the $\pm\vec{Q}$ ordering wave vectors.

3.2 Square lattice Heisenberg antiferromagnet in an external magnetic field

In a ferromagnet, it is relatively easy to study the effect of an external magnetic field. The field breaks the rotational symmetry, and the spins align along the field direction. In an antiferromagnet, it is more complicated. The spins are subjected to two tendencies: to remain antiparallel and to align along the field. Let us see what happens now.

The Hamiltonian is given by

$$H = J\vec{S}_i \cdot \vec{S}_j - h\sum_i S_i^z, \tag{3.9}$$

with $h = g\mu_B B$. We use a rotating frame (x, y, z) such that the local $\hat{z}_i$-axis points in the direction of each magnetic sublattice [3, 4]. As we have seen before, in the absence of a magnetic field, the classical ground state of the antiferromagnet is the Néel state, with wavevector $\vec{Q} = (\pi, \pi)$, in which all spins are aligned antiparallel with respect to one another. We suppose that in the absence of the magnetic field, the spins align in the x–y plane. As the magnetic field is turned on, the model's symmetry is reduced to O(2), describing rotations around the direction of the applied field. The spins align in the plane perpendicular to the magnetic field for small fields, with a small uniform out-of-plane component. That is, the spins cant toward the field direction by an angle θ. As the applied magnetic field is increased, the spins bend round more and more until at the saturation field $h_s = 8JS$ all the spins are aligned ferromagnetically [5]. Note that this is a semiclassical approach since there is no sense saying that a quantum spin makes an angle with the field direction. Also, in a semiclassical picture, we can imagine that at $h = 0$, half of the spins are in the state $|+\rangle$ and the other half in the state $|-\rangle$. Increasing the field, there is a discrete flipping of spins from $|-\rangle$ to $|+\rangle$ until all spins are in the state $|+\rangle$. However, when we compare the semiclassical calculation with numerical simulations, we see that it gives good results.

Spin components in the laboratory frame (x_0, y_0, z_0) are related to those in the local frame by

$$\begin{aligned} S_i^{x_0} &= S_i^x \sin\theta + S_i^z e^{i\vec{Q}\cdot\vec{r}_i} \cos\theta, \quad S_i^{y_0} = S_i^y \\ S_i^{z_0} &= -S_i^x e^{i\vec{Q}\cdot\vec{r}_i} \cos\theta + S_i^z \sin\theta. \end{aligned} \tag{3.10}$$

We get

$$\begin{aligned} S_i^{z_0} S_j^{z_0} + S_i^{x_0} S_j^{x_0} &= (\sin^2\theta - \cos^2\theta)\left(S_i^z S_j^z + S_i^x S_j^x\right) \\ &= -\cos 2\theta\left(S_i^z S_j^z + S_i^x S_j^x\right). \end{aligned} \tag{3.11}$$

Noting that $\sum_i S_i^x e^{i\vec{Q}\cdot\vec{r}_i} = 0$, we can write (3.9) as

$$H = J\sum_{\langle ij\rangle}\left[-\cos 2\theta\left(S_i^x S_j^x + S_i^z S_j^z\right) + S_i^y S_j^y\right] - h\sin\theta\sum_i S_i^z. \tag{3.12}$$

Using the HP representation, we obtain

$$\begin{aligned} &-\cos 2\theta S_i^x S_j^x + S_i^y S_j^y \\ &= \frac{S}{2}\left[-(\cos 2\theta + 1)(a_i a_j + a_i^\dagger a_j^\dagger) + (1 - \cos 2\theta)(a_i a_j^\dagger + a_i^\dagger a_j)\right]. \end{aligned} \tag{3.13}$$

Then the Hamiltonian can be written as

$$\begin{aligned} H_1 = h\sin\theta\sum_i a_i^\dagger a_i + JS\sum_{\langle ij\rangle}&\left[\cos 2\theta\left(a_i^\dagger a_i + a_j^\dagger a_j\right)\right. \\ &\left. + \sin^2\theta\left(a_i^\dagger a_j + a_j^\dagger a_i\right) - \cos^2\theta\left(a_i^\dagger a_j^\dagger + a_j a_i\right)\right] + E_0, \end{aligned} \tag{3.14}$$

where

$$E_0/N = -2S^2J\cos 2\theta - hS\sin\theta. \tag{3.15}$$

Minimization of the classical energy E_0 with respect to the tilting angle θ leads to

$$\sin\theta = h/8JS. \tag{3.16}$$

Fourier transforming the Hamiltonian yields

$$H_1 = \sum_k \left[A_k a_k^\dagger a_k - \frac{1}{2} B_k (a_k a_{-k} + a_k^\dagger a_{-k}^\dagger) \right], \tag{3.17}$$

where

$$A_k = 4JS(1 + \sin^2\theta\gamma_k), \quad B_k = 4JS\cos^2\theta\gamma_k. \tag{3.18}$$

Using the Bogoliubov transformation

$$a_k = u_k\alpha_k + v_k\alpha_{-k}^\dagger, \tag{3.19}$$

we find

$$u_k^2 = \frac{A_k + \varepsilon_k}{2\varepsilon_k}, \quad v_k^2 = \frac{A_k - \varepsilon_k}{2\varepsilon_k}, \tag{3.20}$$

$$\varepsilon_k = \sqrt{A_k^2 - B_k^2} = 4JS\sqrt{(1+\gamma_k)(1-\cos 2\theta\gamma_k)} = 4JS\omega_k, \tag{3.21}$$

and

$$H_1 = \sum_k \varepsilon_k \alpha_k^\dagger \alpha_k + \frac{1}{2}\sum_k (\varepsilon_k - A_k). \tag{3.22}$$

Quantum effects add a zero-point contribution to the ground state energy. The field-induced planar anisotropy opens a gap at $\vec{k} = 0$. The magnetic field produces an effective easy-plane anisotropy. As a consequence, the thermodynamical behavior at fields below the saturation field is similar to the 2d XY-model. We have a BKT transition with the BKT transition temperature depending on the field [6].

3.3 Dzyaloshinskii–Moriya interaction

There are many terms in the spin–spin interaction that we can add to the Heisenberg model [7]. Some are of higher-order or more complicated forms. Here we study the Dzyaloshinskii–Moriya interaction (DMI), which is a term of the form

$$H_{DM} = \vec{D}_{12}.\ (\vec{S}_1 \times \vec{S}_2). \tag{3.23}$$

The DM interaction is antisymmetric since it involves the vector product $\vec{S}_1 \times \vec{S}_2$.

The vector $\vec{D}_{12}$ associated with the bond between $\vec{S}_1$ and $\vec{S}_2$ can be spatially inhomogeneous. The energy of any nearest-neighbor bond on the lattice can be minimized by spin vectors perpendicular to each other and perpendicular to the DM vector $\vec{D}$. In magnetically ordered systems, the DM interaction favors a spin canting of otherwise (anti) parallel aligned spins. We have to analyze the crystal lattice's symmetry operations to see if the term (3.23) is allowed. If a center of inversion is located at the midpoint of the line joining i and j, then $\vec{D}_{12} = 0$. (A center of inversion is a point through the crystal structure, such that if there is an atom at (x, y, z), there must also be an atom at $(-x, -y, -z)$.) If there is a center of inversion, inverting the spins, we obtain $\vec{S}_A \times \vec{S}_B \rightarrow \vec{S}_B \times \vec{S}_A = -\vec{S}_A \times \vec{S}_B$. Thus, the system is invariant only if $\vec{D}_{12} = 0$.

The DM interaction is induced by a lack of inversion symmetry of the compound and strong spin–orbit coupling. That means that we have to know the compound's crystal structure to see if the DMI is allowed. Inversion symmetry can be broken in different directions leading to different induced DM interactions. We also have the following rules:

- When a mirror plane perpendicular to the line ij bisects it, $\vec{D}$ is parallel to the mirror plane.
- When there is a mirror plane including i and j, $\vec{D}$ is perpendicular to the mirror plane.

The nearest-neighbor Heisenberg ferromagnet Hamiltonian with DM interaction is written as

$$H = -J\sum_{\langle ij\rangle}\vec{S}_i \cdot \vec{S}_j + \sum_{\langle ij\rangle}\vec{D}_{ij} \cdot (\vec{S}_i \times \vec{S}_j), \tag{3.24}$$

with $\vec{D}_{ij} = -\vec{D}_{ji}$. If $\vec{D}$ is along the $\hat{z}$ direction we have

$$H = -J\sum_{\langle ij\rangle}\vec{S}_i \cdot \vec{S}_j + D\sum_{\langle ij\rangle}(S_i^x S_j^y - S_i^y S_j^x). \tag{3.25}$$

Let us consider a classical linear chain with the spins in the x–y plane and write

$$S_i^x = \cos\theta_i, \quad S_i^y = \sin\theta_i. \tag{3.26}$$

The first term in (3.25) favors parallel neighboring spin pairs where the second term selects perpendicular spin pairs. (Note that $\hat{e}_z \cdot \vec{a} \times \vec{b} = ab \sin\theta$ is a minimum when $\theta = 3\pi/2$.) We now write (3.26) as

$$H = -\sum_i\left[J\cos(\theta_i - \theta_{i+1}) - D\sin(\theta_i - \theta_{i+1})\right]. \tag{3.27}$$

Let $\phi = \tan^{-1}(D/J)$, with $0 \leqslant \phi \leqslant \pi/2$. Then (3.27) can be written as

$$H = -\sqrt{J^2 + D^2}\sum_i \cos((\theta_i - \theta_{i+1} + \phi)). \tag{3.28}$$

The ground state is given by $\theta_{i+1} - \theta_i = \phi$. That is, the classical ground state has a canted spin structure. Let us now consider the quantum case and write the xy component of H as

$$H_{xy} = -\frac{J}{2}\sum_{\langle ij\rangle}\left(S_i^+S_j^- + S_i^-S_j^+\right) + \frac{iD}{2}\sum_{\langle ij\rangle}\left(S_i^+S_j^- - S_i^-S_j^+\right). \tag{3.29}$$

As before, we take $\phi = \tan^{-1}(D/J)$. Note that

$$\sin\phi = \frac{D}{\sqrt{J^2 + D^2}}, \qquad \cos\phi = \frac{J}{\sqrt{J^2 + D^2}}. \tag{3.30}$$

We can then write

$$H_{xy} = -\frac{1}{2}\sqrt{J^2 + D^2}\sum_{\langle ij\rangle}\left[e^{-i\phi}S_i^+S_j^- + e^{i\phi}S_i^-S_j^+\right]. \tag{3.31}$$

Or more generally

$$H = -\sum_{ij}\left[\frac{\tilde{J}_{ij}}{2}\left(e^{-i\phi_{ij}}S_i^+S_j^- + e^{i\phi_{ij}}S_i^-S_j^+\right) + J_{ij}S_i^zS_j^z\right], \tag{3.32}$$

where $\tilde{J}_{ij}e^{i\phi_{ij}} = J_{ij} + iD_{ij}$.

It is usual to write the MD term as $D\sum_{\langle ij\rangle}\nu_{ij}\hat{z}\cdot(\vec{S}_i \times \vec{S}_j)$, where $\nu_{ij} = -\nu_{ji} = \pm 1$. And so we can calculate the structure factor for the MD term in the same way as we do for γ_k. Hence, in a square lattice, we take 1 for a bond j to $j+1$, and -1 for a bond j to j-1, in the x-direction. We use a similar procedure for the y-direction in a square lattice.

A more general Hamiltonian can be written as

$$H = -\sum_{ij}J_{ij}\vec{S}_i\cdot\vec{S}_j + \sum_{ij}\vec{D}_{ij}\cdot(\vec{S}_i \times \vec{S}_j) + \sum_{ij}\vec{S}_i\cdot\overleftrightarrow{\Gamma}_{ij}\cdot\vec{S}_j. \tag{3.33}$$

The symmetric exchange interaction tensor $\overleftrightarrow{\Gamma}_{ij}$ is called the pseudo-dipolar interaction. This interaction gives an easy-axis anisotropy along the principal axis of the tensor $\overleftrightarrow{\Gamma}_{ij}$. In contrast, the DM interaction gives an easy-plane anisotropy perpendicular to the vector $\vec{D}_{ij}$. For a derivation of the MDI term see [8].

3.4 Symmetries

Space inversion transformation I turns a vector $\vec{r}$ into $-\vec{r}$. It also changes $\vec{k}$ into $-\vec{k}$. If a system is invariant under space inversion, the dispersion relation is an even function of $\vec{k}$: $\omega(\vec{k}) = -\omega(-\vec{k})$.

Time-reversal transformation T changes the direction of time. It does not change the direction of $\vec{r}$, but it changes the direction of the velocity and, therefore, the momentum and $\vec{k}$. So, it changes the direction of the angular momentum and hence the direction of the spin. It turns a complex number into its conjugate (it is an anti-unitary transformation). It also changes the direction of a magnetic field. We can see this as follows. A current flowing in a wire produces a radial magnetic field. If we reverse the time direction, the current will flow in the opposite direction, and hence the magnetic field will also point in the opposite direction.

In the paramagnetic state, the time-reversal operation will invert all spins but leave all observables unchanged. Since the spins point in all directions, the process will leave the system invariant. In contrast, in the ferromagnetic state, where the spins point in a specific direction, the time-reversal transformation would have the effect of inverting the magnetization, so it would have a macroscopically observable impact. The symmetry is broken.

In an antiferromagnet, time-reversal symmetry is also broken since flipping all the spins would have the effect of swapping the A and B sublattices, changing the sign of the staggered magnetization. Also, another symmetry is broken, namely the symmetry under lattice translation.

Kramer's theorem states that all eigenstates of a time-reversal invariant Hamiltonian of spin ½ system are twofold degenerate [9]. The bipartite antiferromagnet is invariant under a combination of time-reversal symmetry and lattice translation of one site. This gives an analog of Kramer's theorem, and consequently, the dispersion relations are doubly degenerate.

3.5 Nonlinear spin-wave theory

The spin-wave theory is a good starting point to describe magnetic ordering. However, we can improve the theory with the inclusions of fourth-order terms. Of course, the procedure can only be applied to magnetically ordered phases. As an example, I will consider the antiferromagnetic square lattice on two dimensions with an easy-axis anisotropy term. The system will then be in an ordered phase below a critical temperature T_N. It is known that magnon–magnon interactions do not play an essential role in an ordered ferromagnet at low temperatures. I will present a self-consistent mean-field method to deal with the nonlinear terms of the Hamiltonian. The procedure can be extended to other models.

The Hamiltonian of my interest is given by

$$H = J\sum_{\langle ij\rangle}\vec{S}_i \cdot \vec{S}_j - A\sum_i (S_i^z)^2. \tag{3.34}$$

Using the HP representation up to four operator terms:

$$S_i^+ = \sqrt{2S}\left(a_i - \frac{a_i^\dagger a_i a_i}{4S}\right), \quad S_i^- = \sqrt{2S}\left(a_i^\dagger - \frac{a_i^\dagger a_i^\dagger a_i}{4S}\right), \quad S_i^z = S - a_i^\dagger a_i, \tag{3.35}$$

on sublattice A, and

$$S_j^+ = \sqrt{2S}\left(b_j^\dagger - \frac{b_j^\dagger b_j^\dagger b_j}{4S}\right), \quad S_j^- = \sqrt{2S}\left(b_j - \frac{b_j^\dagger b_j b_j}{4S}\right), \quad S_j^z = -S + b_j^\dagger b_j, \tag{3.36}$$

on sublattice B, we have

$$S_i^+ S_j^- = 2S\left[a_i b_j - \frac{a_i b_j^\dagger b_j b_j}{4S} - \frac{a_i^\dagger a_i a_i b_j}{4S}\right], \tag{3.37}$$

$$S_i^- S_j^+ = 2S\left[a_i^\dagger b_j^\dagger - \frac{a_i^\dagger b_j^\dagger b_j^\dagger b_j}{4S} - \frac{a_i^\dagger a_i^\dagger a_i b_j^\dagger}{4S}\right]. \tag{3.38}$$

We can write the $S_i^z S_j^z$ term in a symmetric form as

$$\begin{aligned}&\frac{1}{2}\left(S_i^z S_j^z + S_j^z S_i^z\right)\\ &=\frac{1}{2}\left[-2S^2 + Sb_j^\dagger b_j + Sa_i^\dagger a_i + Sb_i^\dagger b_i + Sa_j^\dagger a_j - a_i^\dagger a_i b_j^\dagger b_j - b_j^\dagger b_j a_i^\dagger a_i\right].\end{aligned} \tag{3.39}$$

To get an idea of what happens when we take the Fourier transform, let us consider as an example the term

$$\begin{aligned}T_1 &= \sum_{ij} a_i b_j^\dagger b_j b_j = \frac{1}{2}\sum_r \sum_\delta a_{r+\delta} b_r^\dagger b_r b_r\\ &= \frac{1}{2}\sum_{r,\delta}\sum_{k_1}\sum_{k_2}\sum_{k_3}\sum_{k_4} e^{-i\vec{k}_1\cdot(\vec{r}+\vec{\delta})+i\vec{k}_2\cdot\vec{r}-i\vec{k}_3\cdot\vec{r}-i\vec{k}_4\cdot\vec{r}} a_{k_1} b_{k_2}^\dagger b_{k_3} b_{k_4}.\end{aligned}$$

The term vanishes unless $\vec{k}_2 = \vec{k}_1 + k_3 + k_4$. Hence we have

$$T_1 = \frac{1}{2}\sum_{k_1 k_2 k_3}\sum_\delta e^{-i\vec{k}_1\vec{\delta}} a_{k_1} b_{k_1+k_3+k_4}^\dagger b_{k_3} b_{k_4}. \tag{3.40}$$

Let, as before

$$\gamma(\vec{k}_1) = \sum_\delta e^{i\vec{k}_1\cdot\vec{\delta}}. \tag{3.41}$$

Renaming $\vec{k}_4 \to \vec{k}_2$, we get

$$T_1 = \frac{1}{2}\sum_{k_1 k_2 k_3} \gamma(\vec{k}_1) a_{k_1} b_{k_1+k_2+k_3}^\dagger b_{k_3} b_{k_2}. \tag{3.42}$$

When all the terms are collected, the Hamiltonian becomes

$$\begin{aligned}H_2 &= 2JS\sum_k\left[a_k^\dagger a_k + b_k^\dagger b_k + \gamma(\vec{k})\left(a_k b_{-k} + a_k^\dagger b_{-k}^\dagger\right)\right]\\ &- J\sum_{k_1,k_2,k_3}\left\{\frac{\gamma(\vec{k}_1)}{8}\left(a_{k_1} b_{k_1+k_2+k_3}^\dagger b_{k_2} b_{k_3} + a_{k_1+k_2+k_3}^\dagger a_{k_2} a_{k_3} b_{k_1} + a_{k_1}^\dagger b_{k_2}^\dagger b_{k_3}^\dagger b_{k_1+k_2+k_3}\right.\right.\\ &\left.\left.+ a_{k_2}^\dagger a_{k_3}^\dagger a_{k_1+k_2+k_3} b_{k_1}^\dagger\right) + \frac{\gamma(\vec{k}_2-\vec{k}_1)}{4}\left(a_{k_1}^\dagger a_{k_2} b_{k_3}^\dagger b_{k_1-k_2+k_3} + b_{k_1}^\dagger b_{k_2} a_{k_3}^\dagger a_{k_1-k_2+k_3}\right)\right\}.\end{aligned} \tag{3.43}$$

Now we perform a mean-field decoupling on quartic terms to obtain an effective quadratic Hamiltonian. We use the relation:

$$\begin{aligned}\mathrm{ABCD} = \mathrm{AB}\langle\mathrm{CD}\rangle + \langle\mathrm{AB}\rangle\mathrm{CD} + \mathrm{AC}\langle\mathrm{BD}\rangle + \langle\mathrm{AC}\rangle\mathrm{BD} + \mathrm{AD}\langle\mathrm{BC}\rangle \\ + \langle\mathrm{AD}\rangle\mathrm{BC}.\end{aligned} \tag{3.44}$$

From equation (2.37) and considering that the only non-null terms are

$$\langle\alpha_k^\dagger\alpha_k\rangle,\ \langle\alpha_k\alpha_k^\dagger\rangle,\ \langle\beta_k^\dagger\beta_k\rangle,\ \langle\beta_k\beta_k^\dagger\rangle,$$

we have

$$\langle a_k a_k\rangle = \langle b_k b_k\rangle = \langle a_k^\dagger a_k^\dagger\rangle = \langle b_k^\dagger b_k^\dagger\rangle = 0. \tag{3.45}$$

We can proceed now and do the decoupling. For instance

$$\begin{aligned}&a_{k_1} b^\dagger_{k_1+k_2+k_3} b_{k_3} b_{k_2} \longrightarrow \\ &\langle a_{k_1} b_{k_3}\rangle b^\dagger_{k_1+k_2+k_3} b_{k_2} + a_{k_1} b_{k_3}\langle b^\dagger_{k_1+k_2+k_3} b_{k_2}\rangle \\ &+ \langle a_{k_1} b_{k_2}\rangle b^\dagger_{k_1+k_2+k_3} b_{k_3} + a_{k_1} b_{k_2}\langle b^\dagger_{k_1+k_2+k_3} b_{k_3}\rangle \\ &= \langle a_{k_1} b_{-k_1}\rangle b^\dagger_{k_2} b_{k_2} + a_{k_1} b_{-k_1}\langle b^\dagger_{k_2} b_{k_2}\rangle \\ &+ \langle a_{k_1} b_{-k_1}\rangle b^\dagger_{k_3} b_{k_3} + a_{k_1} b_{-k_1}\langle b^\dagger_{k_3} b_{k_3}\rangle.\end{aligned} \tag{3.46}$$

The four operator term without $S_i^z S_j^z$ is

$$\begin{aligned}H_2' = -\frac{J}{4}\sum_{k_1,k_2}\Big\{&\gamma(\vec{k}_1)[\langle a_{k_2}^\dagger a_{k_2}\rangle + \langle b_{k_2}^\dagger b_{k_2}\rangle]\big(a_{k_1} b_{-k_1} + a_{k_1}^\dagger b_{-k_1}^\dagger\big) \\ &+ \gamma(\vec{k}_2)[\langle a_{k_2} b_{-k_2}\rangle + \langle a_{k_2}^\dagger b_{k_2}^\dagger\rangle]\big(a_{k_1}^\dagger a_{k_1} + b_{k_1}^\dagger b_{k_1}\big)\Big\}.\end{aligned} \tag{3.47}$$

We can write (3.47) as

$$H_2' = J(\Gamma_1 + \Gamma_2)\sum_k\Big\{\gamma(\vec{k})\big(a_k b_{-k} + a_k^\dagger b_{-k}^\dagger\big) + \big(a_k^\dagger a_k + b_k^\dagger b_k\big)\Big\}, \tag{3.48}$$

where

$$\Gamma_1 = \sum_k(\langle a_k^\dagger a_k\rangle + \langle b_k^\dagger b_k\rangle), \quad \Gamma_2 = \sum_k\gamma(\vec{k})(\langle a_k b_{-k}\rangle + \langle a_k^\dagger b_{-k}^\dagger\rangle). \tag{3.49}$$

For the $S_i^z S_j^z$ term we get

$$H_2^z = -J\sum_k\Big[\Gamma_1(a_k^\dagger a_k + b_k^\dagger b_k) + \gamma(\vec{k})\Gamma_2(a_k b_{-k} + a_k^\dagger b_{-k}^\dagger)\Big], \tag{3.50}$$

where I am considering lattices where the relation $\gamma(\vec{k}_2 - \vec{k}_1) = \gamma(\vec{k}_2)\gamma(\vec{k}_1)$ holds.

Let us now consider the term with the anisotropy. We write

$$\begin{aligned} H_A &= -\frac{A}{2}\left[\sum_{n\in A}(S_n^z)^2 + \sum_{n\in B}(S_n^z)^2\right] \\ &= -AS\sum_n(a_n^\dagger a_n + b_n^\dagger b_n) + \frac{A}{2}\sum_n(a_n^\dagger a_n a_n^\dagger a_n + b_n^\dagger b_n b_n^\dagger b_n). \end{aligned} \tag{3.51}$$

Using the commutation relations $[a_n^\dagger, a_n] = 1$, $[b_n^\dagger, b_n] = 1$, we can write (3.51) as

$$H_A = -A\sum_n\left[(2S-1)(a_n^\dagger a_n + b_n^\dagger b_n) + a_n^\dagger a_n^\dagger a_n a_n + b_n^\dagger b_n^\dagger b_n b_n\right]. \tag{3.52}$$

The decoupling of the four operator terms leads to a term

$$4\left(\langle a_{k_2}^\dagger a_{k_2}\rangle a_{k_1}^\dagger a_{k_1} + \langle b_{k_2}^\dagger b_{k_2}\rangle b_{k_1}^\dagger b_{k_1}\right). \tag{3.53}$$

The final result for the mean-field Hamiltonian is then

$$\begin{aligned} H^{MF} &= J\left[2S - (\Gamma_1 + \Gamma_2)\right]\sum_k\left[a_k^\dagger a_k + b_k^\dagger b_k + \gamma(\vec{k})(a_k b_{-k} + a_k^\dagger b_{-k}^\dagger)\right] \\ &\quad - A\left[(2S-1) - \Gamma_1\right]\sum_k(a_k^\dagger a_k + a_k a_k^\dagger + b_k^\dagger b_k + b_k b_k^\dagger). \end{aligned} \tag{3.54}$$

The mean-field Hamiltonian is similar to the quadratic Hamiltonian with renormalized parameters $2S\tilde{J} = J[2S - (\Gamma_1 + \Gamma_2)]$, $\tilde{A}S = A(2S - 1 - \Gamma_1)$.

Using equation (2.37) we find

$$\begin{aligned} \langle a_k^\dagger a_k\rangle &= u_k^2\langle \alpha_k^\dagger \alpha_k\rangle + v_k^2\langle \beta_k\beta_k^\dagger\rangle = u_k^2 n_\alpha + v_k^2(1 + n_\beta), \\ \langle a_k a_k^\dagger\rangle &= u_k^2(1 + n_\alpha) + v_k^2 n_\beta, \\ \langle b_k^\dagger b_k\rangle &= v_k^2(1 + n_\alpha) + u_k^2 n_\beta, \\ \langle b_k b_k^\dagger\rangle &= v_k^2 n_\alpha + u_k^2(1 + n_\beta), \\ \langle a_k b_{-k}\rangle + \langle a_k^\dagger b_{-k}^\dagger\rangle &= 2u_k v_k(1 + n_\alpha + n_\beta), \end{aligned} \tag{3.55}$$

which leads to

$$\Gamma_1 = \sum_k\left[2u_k^2 + (u_k^2 + v_k^2)(n_\alpha + n_\beta)\right], \quad \Gamma_2 = 2\sum_k \gamma(\vec{k})u_k v_k(1 + n_\alpha + n_\beta). \tag{3.56}$$

One has then to solve a set of self-consistent equations to obtain the parameters Γ_1 and Γ_2.

3.6 Modified spin-wave theory

Takahashi [10] has formulated a generalized spin-wave theory for low-dimensional Heisenberg magnets that supplement the usual spin-wave approach for ferromagnets and antiferromagnets with the constraint that the total magnetization in the case of the ferromagnet or the total staggered magnetization in the case of the antiferromagnet is zero (remember that here one has disordered phases). In the following, I present the case of the two-dimensional Heisenberg antiferromagnet on a square lattice. Takahashi has used the Dyson–Maleev (DM) transformation:

$$S_i^- = a_i^\dagger,\ S_i^+ = (2S - a_i^\dagger a_i)a_i,\ S_i^z = S - a_i^\dagger a_i,\ i \in A \tag{3.57}$$

$$S_j^- = -b_j,\ S_j^+ = -b_j^\dagger(2S - b_j^\dagger b_j),\ S_j^z = -S + b_j^\dagger b_j, j \in B. \tag{3.58}$$

As we saw before, the DM representation avoids the square root, and there is no term higher than the fourth-order in the Hamiltonian. However, the Hamiltonian is nonHermitian. The Hamiltonian is written as follows

$$\begin{aligned} H_1 = &-\frac{1}{2}JzS^2N \\ &+ J\sum_{\langle ij\rangle}\left[S(a_i^\dagger a_i + b_j^\dagger b_j - a_i^\dagger b_j^\dagger - a_i b_j) + \frac{1}{2}a_i^\dagger(b_j^\dagger - a_i)^2 b_j\right], \end{aligned} \tag{3.59}$$

where z is the number of nearest neighbors and N is the total number of sites.

In the following, the ideal spin-wave density matrix with Bogoliubov transformation is introduced

$$\rho = \exp\left[\frac{1}{T}\sum_k{}' \varepsilon_k\left(\alpha_k^\dagger\alpha_k + \beta_{-k}^\dagger\beta_{-k}\right)\right], \tag{3.60}$$

where

$$\alpha_k = \cosh\theta_k a_k - \sinh\theta_k b_{-k}^\dagger, \quad \beta_{-k}^\dagger = -\sinh\theta_k a_k + \cosh\theta_k b_{-k}^\dagger. \tag{3.61}$$

Here $\sum_k'$ means the sum of $\vec{k}$ is over half of the Brillouin zone and

$$a_k = \sqrt{\frac{2}{N}}\sum_{i\in A} e^{-i\vec{k}\cdot\vec{r}_i}a_i, \quad b_{-k}^\dagger = \sqrt{\frac{2}{N}}\sum_{j\in B} e^{-i\vec{k}\cdot\vec{r}_j}b_j^\dagger. \tag{3.62}$$

Using the density matrix (3.60) we get

$$\begin{aligned} &\langle a_i^\dagger b_j\rangle = \langle a_i b_j^\dagger\rangle = \langle a_i^\dagger a_{i'}^\dagger\rangle = \langle a_i a_{i'}\rangle = \langle b_j^\dagger b_{j'}^\dagger\rangle = \langle b_j b_{j'}\rangle = 0, \\ &\langle a_i^\dagger a_{i'}\rangle = f(\vec{r}_i - \vec{r}_{i'}) - \frac{1}{2}\delta_{ii'}, \quad \langle b_j^\dagger b_{j'}\rangle = f(\vec{r}_j - \vec{r}_{j'}) - \frac{1}{2}\delta_{jj'} \\ &\langle a_i^\dagger b_j^\dagger\rangle = \langle a_i b_j\rangle = g(\vec{r}_i - \vec{r}_j), \end{aligned} \tag{3.63}$$

$$\begin{aligned} f(\vec{r}) &\equiv \frac{2}{N}\sum_k{}' \cosh(2\theta_k)\exp(-i\vec{k}\cdot\vec{r})\left(n_k + \tfrac{1}{2}\right), \\ g(\vec{r}) &\equiv \frac{2}{N}\sum_k{}' \sinh(2\theta_k)\exp(-i\vec{k}\cdot\vec{r})\left(n_k + \tfrac{1}{2}\right), \end{aligned} \tag{3.64}$$

$$n_k = \langle \alpha_k^\dagger \alpha_k \rangle = \langle \beta_k^\dagger \beta_k \rangle = [\exp(\varepsilon_k / T) - 1]^{-1}. \tag{3.65}$$

Using the MFA decoupling, we find

$$\langle \vec{S}_i \cdot \vec{S}_j \rangle = -\left[S + \frac{1}{2} - f(0) + g(\vec{r}_i - \vec{r}_j)\right]^2, \, i \in A, j \in B, \tag{3.66}$$

$$\langle \vec{S}_i \cdot \vec{S}_{i'} \rangle = \left[S + \frac{1}{2} - f(0) + f(\vec{r}_i - \vec{r}_{i'})\right]^2, \, i, i' \in A \text{ or } i, i' \in B, i \neq i'. \tag{3.67}$$

The magnetization of each site should be zero for $T > 0$:

$$\langle S_i^z \rangle = S + \frac{1}{2} - f(0) = 0. \tag{3.68}$$

The energy is

$$E = -\left(\frac{JN}{2}\right)\sum_{\delta}\left[S + \tfrac{1}{2} - f(0) + g(\vec{\delta})\right]^2, \tag{3.69}$$

where $\vec{\delta}$ are vectors to the nearest neighbors. The entropy is given by

$$2\sum_k{}' [(n_k + 1) \ln (n_k + 1) - n_k \ln n_k]. \tag{3.70}$$

Now, we minimize $F = E—TS$ under the condition (3.68). Using

$$\frac{\partial}{\partial \theta_k}\left[\frac{F}{N} - \mu f(0)\right] = 0, \quad \frac{\partial}{\partial \varepsilon_k}\left[\frac{F}{N} - \mu f(0)\right] = 0, \tag{3.71}$$

we get

$$\tanh(2\theta_k) = \eta\gamma_k, \, \varepsilon_k = \lambda[\cosh(2\theta_k) - \eta\gamma_k \sinh(2\theta_k)], \quad \lambda \equiv Jzg(\vec{\delta}) - \mu, \tag{3.72}$$

$$\eta \equiv Jzg(\vec{\delta})/\lambda, \quad \gamma_k = z^{-1}\sum_{\delta} e^{i\vec{k} \cdot \vec{\delta}}, \tag{3.73}$$

where μ is a Lagrange multiplier. From equation (3.71) we obtain

$$\varepsilon_k = \lambda\left(1 - \eta^2\gamma_k^2\right)^{1/2}, \quad \cosh(2\theta_k) = \frac{1}{\left(1 - \eta^2\gamma_k^2\right)^{1/2}}, \quad \sinh(2\theta_k) = \frac{\eta\gamma_k}{\left(1 - \eta^2\gamma_k^2\right)^{1/2}}. \tag{3.74}$$

From (3.67), (3.71), (3.72) and (3.73) we get the self-consistent equations for η and λ,

$$S + \frac{1}{2} = \frac{2}{N}{\sum_k}' \frac{1}{2\left(1 - \eta^2\gamma_k^2\right)^{1/2}} \coth\left[\frac{\lambda\left(1 - \eta^2\gamma_k^2\right)^{1/2}}{2T}\right], \tag{3.75}$$

$$\frac{\eta\lambda}{Jz} = \frac{2}{N}{\sum_k}' \frac{\eta\gamma_k^2}{2\left(1 - \eta^2\gamma_k^2\right)^{1/2}} \coth\left[\frac{\lambda\left(1 - \eta^2\gamma_k^2\right)^{1/2}}{2T}\right]. \tag{3.76}$$

3.7 Exercises

3.1. Solve the triangular antiferromagnetic lattice with a magnetic field in the lattice plane.

3.2. Use the nonlinear spin-wave technique to study the ferromagnetic lattice in three dimensions in the ordered phase.

References

[1] Chernyshev A L and Zhitomirsky M E 2009 Spin-waves in triangular lattice antiferromagnet: decays, spectrum renormalization, and singularities *Phys. Rev.* B **79** 144416

[2] Merdan M and Xian Y 2012 Longitudinal excitations in triangular lattice antiferromagnets *J. Low. Temp Phys.* **171** 797–807

[3] Zhitomirsky M E and Nikuni T 1997 Two-dimensional Heisenberg antiferromagnet in strong magnetic fields *Physica* B **241** 573–5

[4] Mourigal M, Zhitomirsky M E and Chernyshev A L 2010 Field-induced decay dynamics in square-lattice antiferromagnets *Phys. Rev.* B **82** 144402

[5] Blundel S 2003 *Magnetism in Condensed Matter* (Oxford: Oxford University Press)

[6] Landau D P and Binder K 1981 Phase diagrams and critical behavior of a two-dimensional anisotropic Heisenberg antiferromagnet *Phys. Rev.* B **24** 1391–403

[7] Stancel D D and Prabhakar A 2009 *Spin Waves Theory and Applications* (Berlin: Springer)

[8] Moriya T 1960 Anisotropic superexchange interaction and weak ferromagnetism *Phys. Rev.* **120** 91–8

[9] Kittel C 1963 *Quantum Theory of Solids* (New York: Wiley)

[10] Takahashi M 1989 Modified spin-wave theory of a square-lattice antiferromagnet *Phys. Rev.* B **40** 2494–501

Chapter 4

Lattices with two inequivalent sites

In this chapter, we will study more complicated lattices in two dimensions with two inequivalent sites. That is different from the antiferromagnet on a square lattice studied before, where all the crystal sites were equivalent, and only the spin orientations were different on sublattices A and B. We will also see that we will need a generalized Bogoliubov transformation to study the antiferromagnet on these lattices. We start with the honeycomb lattice shown in figure 4.1. As we can see, there are two nonequivalent sites A and B. The three nearest-neighbor vectors are given by (setting the lattice constant as unity)

$$\vec{\delta}_1 = (1, \sqrt{3})/2, \quad \vec{\delta}_2 = (1, -\sqrt{3})/2, \quad \vec{\delta}_3 = (-1, 0).$$

Performing the same steps as we did for the triangular lattice in chapter 3, we find that the structure factor $\gamma_k = \sum_i e^{i\vec{k}\cdot\vec{\delta}_i}/z$ (where $z = 3$) is given by

$$\gamma_k = \frac{1}{3}\left[e^{ik_x} + 2e^{-ik_x/2}\cos\left(\frac{\sqrt{3}}{2}k_y\right)\right]. \tag{4.1}$$

Note that γ_k is complex on the honeycomb lattice.

4.1 The ferromagnetic honeycomb lattice

Since we have two nonequivalent sites, we have to use two kinds of bosonic operators a and b. However, as the spins point in the same direction on both sublattices, the Holstein–Primakoff representation has the same form. That is

$$\begin{aligned} S_i^+ &= \sqrt{2S}\,a_i, \;\; S_i^- = \sqrt{2S}\,a_i^\dagger, \;\; S_i^z = S - a_i^\dagger a_i, \;\; \text{on site } A. \\ S_j^+ &= \sqrt{2S}\,b_j, \;\; S_j^- = \sqrt{2S}\,b_j^\dagger, \;\; S_j^z = S - b_j^\dagger b_j, \;\; \text{on site } B. \end{aligned} \tag{4.2}$$

doi:10.1088/978-0-7503-3879-0ch4

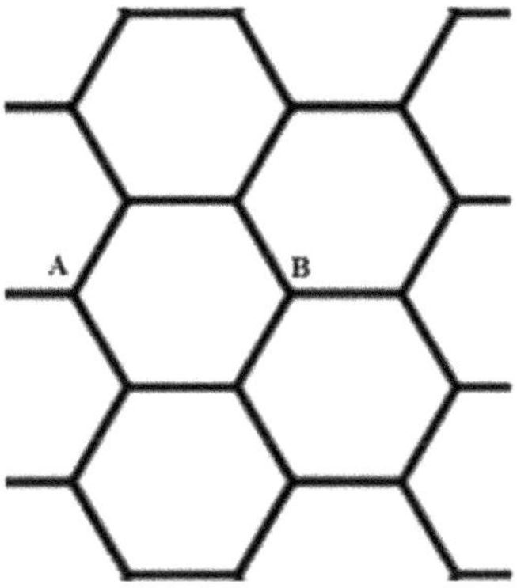

Figure 4.1. The honeycomb lattice.

The Fourier-transformed Hamiltonian assumes the form

$$H_1 = 3JS\sum_k (a_k^\dagger a_k + b_k^\dagger b_k) + \sum_k \left[(h_x - ih_y)a_k^\dagger b_k + (h_x + ih_y)b_k^\dagger a_k\right], \tag{4.3}$$

where I have decomposed γ_k in real and imaginary components and written $JS\gamma_k = h_x + ih_y$. Hamiltonian (4.3) can be written as

$$H_1 = \sum_k (a_k^\dagger, b_k^\dagger)\begin{pmatrix} 3JS & h_x - ih_y \\ h_x + ih_y & 3JS \end{pmatrix}\begin{pmatrix} a_k \\ b_k \end{pmatrix}. \tag{4.4}$$

Introducing $\psi_k^\dagger = (a_k^\dagger, b_k^\dagger)$ we can write (4.4) as

$$H_1 = \sum_k \psi_k^\dagger (3JSI + h_x\sigma_x + h_y\sigma_y)\psi_k, \tag{4.5}$$

where

$$I = \begin{pmatrix} 1 & 0 \\ 0 & 1 \end{pmatrix}, \quad \sigma_x = \begin{pmatrix} 0 & 1 \\ 1 & 0 \end{pmatrix}, \quad \sigma_y = \begin{pmatrix} 0 & -i \\ i & 0 \end{pmatrix}. \tag{4.6}$$

Hamiltonian (4.5) is a particular example of a more general case where we have

$$H_1 = \sum_k \psi_k^\dagger (Ih_0 + \vec{h} \cdot \vec{\sigma})\psi_k, \tag{4.7}$$

with

$$\sigma_z = \begin{pmatrix} 1 & 0 \\ 0 & -1 \end{pmatrix}. \tag{4.8}$$

Hence, I will study Hamiltonian (4.7). Diagonalizing the matrix

$$M_k = \vec{h} \cdot \vec{\sigma} = \begin{pmatrix} h_z & h_x - ih_y \\ h_x + ih_y & -h_z \end{pmatrix}, \tag{4.9}$$

we find the eigenvalues

$$\lambda_{\pm} = \pm h(\vec{k}) = \pm\sqrt{h_x^2 + h_y^2 + h_z^2}, \tag{4.10}$$

and so the dispersion relations are given by

$$\omega(\vec{k}) = h_0 \pm h(\vec{k}). \tag{4.11}$$

For Hamiltonian (4.5), equation (4.11) reduces to

$$\omega_{\pm}(\vec{k}) = 3JS(1 \pm |\gamma_k|). \tag{4.12}$$

To calculate the eigenvectors of M_k we write

$$\psi_k = C_k\varphi_k, \tag{4.13}$$

where

$$\varphi_k = \begin{pmatrix} \alpha_k \\ \beta_k \end{pmatrix}. \tag{4.14}$$

After a straightforward calculation, we find that the eigenvectors are

$$|\vec{k}, +\rangle = \begin{pmatrix} p_1 e^{-i\phi} \\ p_2 \end{pmatrix}, \quad |\vec{k}, -\rangle = \begin{pmatrix} p_2 e^{-i\phi} \\ -p_1 \end{pmatrix}, \tag{4.15}$$

where

$$p_1 = \sqrt{\frac{h + h_z}{2h}}, \quad p_2 = \sqrt{\frac{h - h_z}{2h}}, \quad \tan\phi = \frac{h_y}{h_x}. \tag{4.16}$$

Hence, we can write the matrix C_k as

$$C_k = \begin{pmatrix} p_1 e^{-i\phi} & p_2 e^{-i\phi} \\ p_2 & -p_1 \end{pmatrix}. \tag{4.17}$$

Note that we have the relations

$$\langle\vec{k}, +|\vec{k}, +\rangle = \langle\vec{k}, -|\vec{k}, -\rangle = p_1^2 + p_2^2 = 1, \ \langle\vec{k}, +|\vec{k}, -\rangle = 0. \tag{4.18}$$

4.2 Generalized Bogoliubov transformation

Here I present a formalism to diagonalize more general Hamiltonians, which appears when we study antiferromagnetic models. This can be considered a generalized Bogoliubov canonical transformation, and for this purpose, I consider a general quadratic bosonic Hamiltonian written as

$$H_1 = \sum_k \psi_k^\dagger M_k \psi_k, \tag{4.19}$$

where

$$\psi_k^\dagger = (b_{1k}^\dagger, \ldots, b_{Nk}^\dagger, b_{1k}, \ldots, b_{Nk}). \tag{4.20}$$

Let $\alpha_{mk}(\alpha^{\dagger}_{mk})$ be the eigenmodes of M_k, and write

$$\psi_k = T\varphi_k, \quad \varphi^{\dagger}_k \equiv (\alpha^{\dagger}_{1k}, \ldots, \alpha^{\dagger}_{Nk}, \alpha_{1k}, \ldots, \alpha_{Nk}). \tag{4.21}$$

The standard bosonic commutation relations for the ψ_k and φ_k fields can be written in a matrix form as

$$\left[\psi_{ik}, \psi^{\dagger}_{jk}\right] = \left[\varphi_{ik}, \varphi^{\dagger}_{jk}\right] = \eta_{ij}, \tag{4.22}$$

where

$$\eta \equiv \begin{pmatrix} I_{N\times N} & 0 \\ 0 & -I_{N\times N} \end{pmatrix}. \tag{4.23}$$

Here $I_{N\times N}$ is a $N \times N$ unit matrix. We must ensure that the commutation relations of the bosonic operators must be conserved by the transformation T. The matrix T is chosen such that the Hamiltonian (4.19) can be written as

$$H_1 = \sum_k \varphi^{\dagger}_k T^{\dagger} M_k T \varphi_k, \tag{4.24}$$

where

$$T^{\dagger} M_k T = \begin{pmatrix} \omega_{1k} & 0 & .. & 0 \\ 0 & \omega_{2k} & .. & 0 \\ \ldots & \ldots\ldots & \ldots & \ldots \\ 0 & 0 & \ldots. & \omega_{2Nk} \end{pmatrix}. \tag{4.25}$$

To ensure that the new operators $\alpha^{\dagger}_{mk}$ and α_{mk} satisfy the bosonic algebra, the matrix T must fulfill the condition

$$T\eta T^{\dagger} = \eta. \tag{4.26}$$

A matrix T satisfying condition (4.26) is referred to as paraunitary. The eigenvalues of M_k are then obtained by diagonalization of the matrix

$$K_k = \eta M_k. \tag{4.27}$$

T consists of all the eigenvectors of K_k.

$$T = [V(\omega_1), \ldots, V(\omega_N), V(-\omega_1), \ldots, V(-\omega_N)], \tag{4.28}$$

with the eigenvectors of V ordered as

$$V^{\dagger}(\omega_i)\eta V(\omega_i) = 1, \quad V^{\dagger}(-\omega_i)\eta V(-\omega_i) = -1, \tag{4.29}$$

for each set $(V(\omega_i), V(-\omega_i))$. We get

$$T^{-1} K_k T = \mathrm{diag}[\omega_1, \ldots, \omega_N, -\omega_1, \ldots, -\omega_N], \tag{4.30}$$

$$T^{-1} M_k T = \mathrm{diag}[\omega_1, \ldots, \omega_N, \omega_1, \ldots, \omega_N]. \tag{4.31}$$

All the eigenvalues of K_k are real and appear in pairs. For more details see [1, 2].

In general, we should write η as

$$\eta = \begin{pmatrix} \sigma_1 & & & \\ & \sigma_2 & & \\ & & \cdot & \\ & & & \sigma_{2N} \end{pmatrix} \tag{4.32}$$

where $\sigma_i = +1$ if the correspondent operator is $b^\dagger$ and $\sigma_i = -1$ if it is b.

As an example, let us consider the antiferromagnet on a square lattice studied in chapter 2. Hamiltonian (2.31) can be written as

$$H_1 = JS\sum_k (a_k, b_k^\dagger)\begin{pmatrix} 1 & \gamma_k \\ \gamma_k & 1 \end{pmatrix}\begin{pmatrix} a_k^\dagger \\ b_k \end{pmatrix}. \tag{4.33}$$

The matrix η is therefore

$$\eta = \begin{pmatrix} -1 & 0 \\ 0 & 1 \end{pmatrix}. \tag{4.34}$$

Hence

$$\eta M = \begin{pmatrix} -1 & -\gamma_k \\ \gamma_k & 1 \end{pmatrix}. \tag{4.35}$$

Diagonalizing the matrix (4.35) we obtain the eigenvalues

$$\lambda_k = \pm\sqrt{1 - \gamma_k^2}, \qquad \omega_k = JSz\lambda_k. \tag{4.36}$$

Calling the eigenvectors u_k and v_k we write

$$T = \begin{pmatrix} u_k & v_k \\ v_k & u_k \end{pmatrix}. \tag{4.37}$$

Calculating u_k and v_k we get the Bogoliubov transformation (2.32). Hence, the standard Bogoliubov transformation is a particular case of a more general transformation. In the following, I present a calculation of the checkerboard lattice's eigenvectors that can be applied to generic two-level systems.

4.3 The antiferromagnetic checkerboard lattice

The checkerboard lattice is shown in figure 4.2. The Hamiltonian is

$$H = J_1\sum_{\langle i,j\rangle} \vec{S}_i \cdot \vec{S}_j + J_2 \sum_{\langle\langle i,j\rangle\rangle} \vec{S}_i \cdot \vec{S}_j. \tag{4.38}$$

The lattice has two nonequivalent sites A and B. We denote the near-neighbor spin interaction J_1 between spins at sites A and B by $\langle i, j\rangle$. $\langle\langle i, j\rangle\rangle$ indicates next near-neighbor interactions J_2 between spins at sites A and A, and B and B. As usual, we take the Nèel order perpendicular to the lattice plane, i. e., spins on the A and B sublattices satisfy

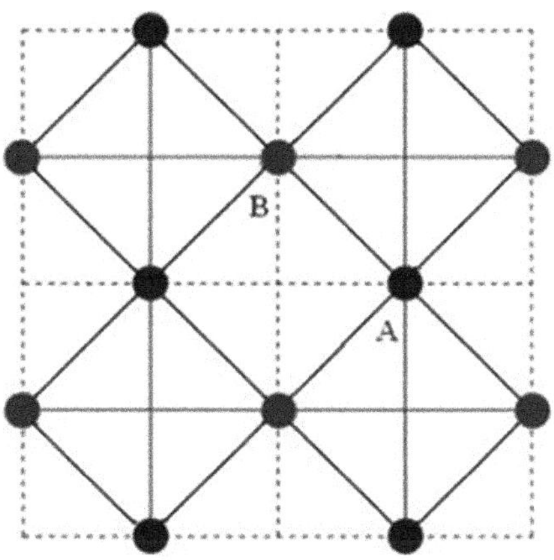

Figure 4.2. The checkerboard lattice.

$\vec{S}_A = -\vec{S}_B = S\hat{z}$ in the ground state. Now we use the linearized HP representation for antiferromagnets

$$\begin{aligned} S_i^+ &= \sqrt{2S}a_i, \ S_i^- = \sqrt{2S}a_i^\dagger, \ S_i^z = S - a_i^\dagger a_i, \ i \in A \\ S_j^+ &= \sqrt{2S}b_j^\dagger, \ S_j^- = \sqrt{2S}b_j, \ S_j^z = -S + b_j^\dagger b_j, \quad j \in B. \end{aligned} \tag{4.39}$$

Taking (4.39) into the Hamiltonian (4.38) and Fourier transforming (using equation (2.32)), we obtain (after discarding the zero-point energy)

$$\begin{aligned} H_1 = \sum_k \Big\{ & g_k(a_k^\dagger a_k + a_k a_k^\dagger) + f_k(b_k^\dagger b_k + b_k b_k^\dagger) \\ & + h_k(a_k b_{-k} + b_k a_{-k} + a_k^\dagger b_{-k}^\dagger + b_k^\dagger a_{-k}^\dagger) \Big\}, \end{aligned} \tag{4.40}$$

where

$$\begin{aligned} g_k &= S[2J_1 + J_2(\cos k_y - 1)], \quad f_k = S[2J_1 + J_2(\cos k_x - 1)], \\ h_k &= 2J_1 S\gamma_k, \end{aligned} \tag{4.41}$$

with

$$\gamma_k = \cos\frac{k_x}{2}\cos\frac{k_y}{2}. \tag{4.42}$$

We can write H_1 as

$$H_1 = \sum_k \psi_k^\dagger H_k \psi_k. \tag{4.43}$$

For this problem, it is more convenient to write $\psi_k^\dagger$ as

$$\psi_k^\dagger = (a_k^\dagger, b_{-k}, a_{-k}, b_k^\dagger). \tag{4.44}$$

With this definition of $\psi_k^\dagger$ we find that the Hamiltonian can be written as two diagonal blocks

$$H_k = \begin{pmatrix} M_k & 0 \\ 0 & M_k \end{pmatrix}, \quad M_k = \begin{pmatrix} g_k & h_k \\ h_k & f_k \end{pmatrix}. \tag{4.45}$$

The matrix η is now given by

$$\eta = \text{diag}(+1, -1, -1, +1). \tag{4.46}$$

Let us consider the first block in (4.45), and write

$$H_1 = \sum_k (a_k^\dagger, b_{-k}) M_k \begin{pmatrix} a_k \\ b_{-k}^\dagger \end{pmatrix}. \tag{4.47}$$

We have

$$\eta = \begin{pmatrix} 1 & 0 \\ 0 & -1 \end{pmatrix} = \sigma_z. \tag{4.48}$$

Diagonalizing

$$\eta M_k = \begin{pmatrix} g_k & h_k \\ -h_k & -f_k \end{pmatrix}, \tag{4.49}$$

we obtain

$$\omega_{\alpha,\pm} = \Delta \pm \sqrt{R^2 - h_k^2}, \tag{4.50}$$

where

$$\Delta = \frac{g_k - f_k}{2}, \quad R = \frac{g_k + f_k}{2}. \tag{4.51}$$

For the second block

$$\eta = \begin{pmatrix} -1 & 0 \\ 0 & 1 \end{pmatrix}, \tag{4.52}$$

and we get

$$\omega_{\beta,\pm} = -\Delta \mp \sqrt{R^2 - h_k^2}. \tag{4.53}$$

We see that

$$\omega_{\beta,\pm} = -\omega_{\alpha,\pm}. \tag{4.54}$$

In the following, we explicitly show the calculation of the eigenvectors [3]. We start with $(\eta M)U = \varepsilon U$, written as

$$\begin{pmatrix} g_k & h_k \\ -h_k & -f_k \end{pmatrix}\begin{pmatrix} x \\ y \end{pmatrix} = \varepsilon\begin{pmatrix} x \\ y \end{pmatrix}. \tag{4.55}$$

We get

$$g_k x + h_k y = \varepsilon x, \quad -h_k x - f_k y = \varepsilon y. \tag{4.56}$$

From (4.56) we have $y = -h_k x/(\varepsilon + f_k)$. Using $x^2 - y^2 = 1$, we find

$$x = \frac{\varepsilon + f_k}{\sqrt{(\varepsilon + f_k)^2 - h_k^2}}. \tag{4.57}$$

The other term is

$$\frac{h_k}{\sqrt{(\varepsilon + f_k)^2 - h_k^2}}. \tag{4.58}$$

The final result is

$$u_\alpha = \frac{h_k}{\sqrt{h_k^2 - (\varepsilon_\alpha - f_k)^2}}, \quad v_\alpha = \frac{f_k - \varepsilon_\alpha}{\sqrt{h_k^2 - (\varepsilon_\alpha - f_k)^2}}, \tag{4.59}$$

$$u_\beta = \frac{\varepsilon_\beta + f_k}{\sqrt{(\varepsilon_\beta + f_k)^2 - h_k}}, \quad v_\beta = \frac{h_k}{\sqrt{(\varepsilon_\beta + f_k)^2 - h_k}}. \tag{4.60}$$

Let us first consider u_β. We have

$$\varepsilon_\beta + f_k = \frac{s}{2} + \frac{\sqrt{s^2 - 4h_k^2}}{2}, \tag{4.61}$$

where $s = f + g$.

$$(\varepsilon_\beta + f_k)^2 - h_k^2 = \sqrt{s^2 - 4h_k^2}(\varepsilon_\beta + f_k),$$

and so

$$u_\beta^2 = \frac{(\varepsilon_\beta + f_k)}{\sqrt{s^2 - 4h_k^2}} = \frac{1}{2}\frac{s + \sqrt{s^2 - 4h_k^2}}{\sqrt{s^2 - 4h_k^2}}. \tag{4.62}$$

Defining

$$\cosh\theta = \frac{s}{\sqrt{s^2 - 4h_k^2}}, \tag{4.63}$$

we get

$$\cosh^2\frac{\theta}{2} = \frac{1}{2}(\cosh\theta + 1) = \frac{1}{2}\left[\frac{s + \sqrt{s^2 - 4h_k^2}}{\sqrt{s^2 - 4h_k^2}}\right]. \tag{4.64}$$

Hence,

$$u_\beta = \cosh\frac{\theta}{2}, \quad \text{and consequently} \quad v_\beta = \sinh\frac{\theta}{2}. \tag{4.65}$$

Following the same procedure, we find

$$\varepsilon_\alpha - f_k = \frac{1}{2}\Big(-s + \sqrt{s^2 - 4h_k^2}\Big), \tag{4.66}$$

$$h_k^2 - (\varepsilon_\alpha - f_k)^2 = \frac{1}{2}\sqrt{s^2 - 4h_k^2}\Big(-\sqrt{s^2 - 4h_k^2} + s\Big), \tag{4.67}$$

$$v_\alpha^2 = \frac{1}{2}\frac{s - \sqrt{s^2 - 4h_k^2}}{\sqrt{s^2 - 4h_k^2}} \rightarrow v_\alpha = \sinh\frac{\theta}{2}, \tag{4.68}$$

and thus $u_\alpha = \cosh\frac{\theta}{2}$.

We can write the final result as

$$\psi_+^\alpha = \begin{pmatrix} \cosh(\theta/2) \\ -\sinh(\theta/2) \\ 0 \\ 0 \end{pmatrix}, \quad \psi_-^\alpha = \begin{pmatrix} -\sinh(\theta/2) \\ \cosh(\theta/2) \\ 0 \\ 0 \end{pmatrix}, \quad \psi_+^\beta = \begin{pmatrix} 0 \\ 0 \\ \cosh(\theta/2) \\ -\sinh(\theta/2) \end{pmatrix}$$
$$\psi_-^\beta = \begin{pmatrix} 0 \\ 0 \\ -\sinh(\theta/2) \\ \cosh(\theta/2) \end{pmatrix}. \tag{4.69}$$

Note that we have

$$\left\langle \psi_\pm^\alpha \middle| \eta \middle| \psi_\pm^\alpha \right\rangle = 1, \; \left\langle \psi_\pm^\alpha \middle| \eta \middle| \psi_\mp^\alpha \right\rangle = 0, \; \left\langle \psi_\pm^\beta \middle| \eta \middle| \psi_\pm^\beta \right\rangle = 1, \; \left\langle \psi_\pm^\beta \middle| \eta \middle| \psi_\mp^\beta \right\rangle = 0. \tag{4.70}$$

4.4 Antiferromagnetic honeycomb lattice

Following the same procedure for the honeycomb lattice as we did for the checkerboard lattice, we obtain now

$$H_k = \begin{pmatrix} H_{kI} & 0 \\ 0 & H_{kII} \end{pmatrix}, \tag{4.71}$$

where

$$H_{kI} = 3JS\begin{pmatrix} 1 & \gamma_k \\ \gamma_k^* & 1 \end{pmatrix}, \quad H_{kII} = H_{-kI}. \tag{4.72}$$

Let us write (4.72) in a generalized form

$$H_{kI} = \begin{pmatrix} g_k & h_k^* \\ h_k & f_k \end{pmatrix}. \tag{4.73}$$

It has the same form as (4.45), except that now h_k is a complex function. We can redo all the calculations of section 4.3, considering h_k a complex function and writing $e^{i\phi}|h_k|$. Doing so, we get for the first block

$$\omega_k = \Delta \pm \sqrt{R^2 - |h_k|^2}. \tag{4.74}$$

For the antiferromagnetic honeycomb lattice, we have then

$$\omega_k = \pm 3JS\sqrt{1 - |\gamma_k|^2}. \tag{4.75}$$

The eigenvectors of the first block, corresponding to the α mode are

$$U_+ = \begin{pmatrix} u_k \\ -v_k \end{pmatrix}, \quad \text{and} \quad U_- = \begin{pmatrix} -v_k \\ u_k^* \end{pmatrix}, \tag{4.76}$$

where

$$u_k = e^{i\phi_k}\cosh\left(\frac{\theta_k}{2}\right), \quad v_k = \sinh\left(\frac{\theta_k}{2}\right), \tag{4.77}$$

$$\tan\phi_k = \frac{\text{Im}\,\gamma_k}{\text{Re}\,\gamma_k}, \quad \cosh\theta_k = \frac{1}{\sqrt{1 - |\gamma_k|^2}}, \quad \sinh\theta_k = \frac{|\gamma_k|}{\sqrt{1 - |\gamma_k|^2}}. \tag{4.78}$$

The eigenvectors (4.76) satisfy the generalized orthonormal conditions

$$\langle U_\pm|\sigma_z|U_\pm\rangle = \pm 1, \quad \text{and} \quad \langle U_\pm|\sigma_z|U_\mp\rangle = 0. \tag{4.79}$$

The eigenvectors are invariant under a U(1) gauge transformation, and hence the position of the term $e^{i\phi_k}$ in the eigenvalues can vary from one author to another.

4.5 The antiferromagnetic Union Jack lattice

The Union Jack lattice [4] is shown in figure 4.3. As we can see, the lattice is translationally invariant when it is displaced along the diagonal. It is the shortest distance as compared with translation along the horizontal or vertical axes. We set each sublattice's spacing equal to 1 and take the components of momentum k_x and k_y along the diagonal axes of the two sublattices.

The Hamiltonian for the antiferromagnetic Union Jack lattice is written as

$$H = J_1\sum_{\langle ij\rangle}\vec{S}_i \cdot \vec{S}_j + J_2\sum_{\langle\langle ij\rangle\rangle}\vec{S}_i \cdot \vec{S}_j. \tag{4.80}$$

The nearest-neighbor AF interaction J_1 connects sites AB. The next nearest-neighbor interaction J_2 (along the diagonals) connects sites AA (see figure 4.3).

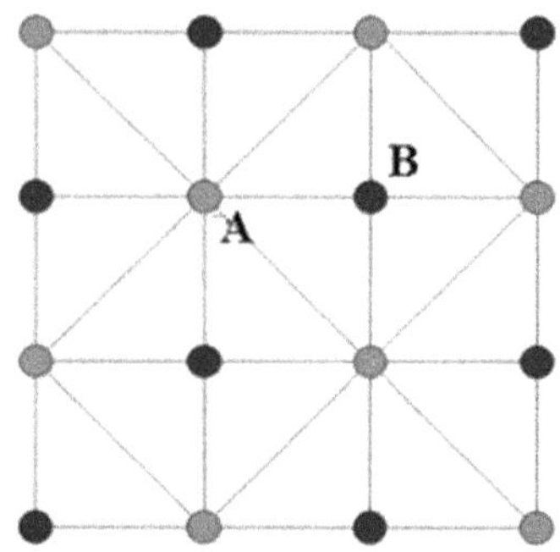

Figure 4.3. The Union Jack lattice. The nearest-neighbor AF interaction J_1 connects sites AB. The next nearest-neighbor interaction J_2 (along the diagonals) connects sites AA.

Using the Holstein–Primakoff transformation and performing a Fourier transform, we get

$$H = 2S\sum_k \left\{ (J_1 - J_2)a_k^\dagger a_k + J_1 b_k^\dagger b_k + J_1\gamma_k(a_k b_{-k} + a_k^\dagger b_{-k}^\dagger) + J_2\eta_k a_k^\dagger a_k \right\}, \quad (4.81)$$

where

$$\gamma_k = \cos\frac{k_x}{2}\cos\frac{k_y}{2}, \quad \eta_k = \frac{\cos k_x + \cos k_y}{2}. \quad (4.82)$$

We can now use the procedure of section (4.3) with

$$g_k = J_1 + J_2(\eta_k - 1), \; f_k = 1, \; h_k = \gamma_k, \quad (4.83)$$

and get the eigenvalues and eigenvectors.

4.6 Exercises

4.1. Consider the magnon–phonon Hamiltonian

$$H = \sum_k \left\{ \omega_k^m a_k^\dagger a_k + \omega_k^p b_k^\dagger b_k + f_k(a_k b_k^\dagger + a_k^\dagger a_k) \right\},$$

where f_k is a coupling constant, $a^\dagger$, a ; $b^\dagger$, b are magnon and phonon creation and annihilation operators. Write the quadratic Hamiltonian in terms of operators $\alpha^\dagger$, α; $\beta^\dagger$, β.

References

[1] Colpa J H P 1978 Diagonalization of the quadratic boson Hamiltonians *Physica* A **93** 327–53

[2] Samajdar R, Chatterjee S, Sachdev S and Scheurer M S 2019 Thermal Hall effect in square lattice spin liquids: a Schwinger boson mean-field study *Phys. Rev.* B **99** 165126

[3] Canals B 2002 From the square lattice to the checkerboard lattice: spin-wave and large-*n* limit analysis *Phys. Rev.* B **65** 184408

[4] Collins A, McEvoy J, Robinson D, Hamer C J and Weihong Z 2006 Quantum spin model with frustration on the union jack lattice *Phys. Rev.* B **73** 024407

IOP Publishing

Theoretical Tools for Spin Models in Magnetic Systems

Antonio Sergio Teixeira Pires

Chapter 5

Schwinger bosons

The spin-wave formalism is very useful when we have a magnetically ordered ground state. However, there are a lot of circumstances when this is not the case. For instance, when we want to treat quantum states which are rotationally symmetric. In these cases, one can use the parton formalism (which does not impose any preferred spin direction) where the spin operators are decomposed into two (or more) kinds of partons that can be either bosonic or fermionic. In this chapter, I start studying the bosonic operators called Schwinger bosons. One of the advantages of the boson formalism over the fermion is that it can give a ground state with Néel order. I also will briefly discuss the fermion operator formalism and study valence bond lattices.

5.1 Schwinger bosons

The Schwinger boson technique can describe different ground-state structures in a unified formalism, without biasing the calculation toward the existence or non-existence of long-range magnetic order. That is, we have a formalism capable of describing any possible ground state. In this approach, there is no need to partition the lattice into different sublattices (although it can be done) or perform any spin operator rotation to a local reference axis, as is customarily done in the standard spin-wave theory.

If $|0\rangle$ represents the Schwinger bosons vacuum, we introduce two operators a and b by

$$|\uparrow\rangle = a^\dagger|0\rangle, \quad |\downarrow\rangle = b^\dagger|0\rangle, \tag{5.1}$$

where $|\uparrow\rangle$ and $|\downarrow\rangle$ are spin-1/2 states. The operators a and b satisfy the standard bosonic commutation relations: $[a, a^\dagger] = 1, [b, b^\dagger] = 1$. We have also

$$\begin{aligned} a|0\rangle = 0, \quad b|0\rangle = 0, \quad a^\dagger|\uparrow\rangle = 0, \\ a|\uparrow\rangle = |0\rangle, \quad b|\downarrow\rangle = |0\rangle, \quad b^\dagger|\downarrow\rangle = 0, \end{aligned} \tag{5.2}$$

doi:10.1088/978-0-7503-3879-0ch5

$a^\dagger$ and $b^\dagger$ create spin ½ particles called spinons. The spin operators are represented by

$$S^+ = S^x + iS^y = a^\dagger b, \qquad S^- = S^x - iS^y = b^\dagger a, \tag{5.3}$$

$$S^z = \frac{1}{2}(a^\dagger a - b^\dagger b). \tag{5.4}$$

We can verify that the commutation relations are satisfied:

$$\begin{aligned}
[S^x, S^y] &= \frac{i}{2}[S^+, S^-] = \frac{i}{2}[a^\dagger b, b^\dagger a] = \frac{i}{2}(a^\dagger a - b^\dagger b) = iS^z \\
[S^z, S^x] &= \frac{1}{2}[S^z, S^+ + S^-] = \frac{1}{4}[a^\dagger a - b^\dagger b, a^\dagger b + b^\dagger a] \\
&= \frac{1}{2}(a^\dagger b - b^\dagger a) = iS^y, \\
[S^y, S^z] &= \frac{1}{2i}[S^+ - S^-, S^z] = \frac{1}{4i}[a^\dagger b - b^\dagger a, a^\dagger a - b^\dagger b] \\
&= \frac{i}{2}(a^\dagger b + b^\dagger a) = iS^x.
\end{aligned} \tag{5.5}$$

We also have

$$\begin{aligned}
S(S+1) &= S^x S^x + S^y S^y + S^z S^z = \frac{1}{2}(S^+S^- + S^-S^+) + S^z S^z \\
&= \frac{1}{2}[(a^\dagger b)^2 + (b^\dagger a)^2] + \frac{1}{4}(a^\dagger a - b^\dagger b)^2 \\
&= \frac{1}{2}(a^\dagger a + b^\dagger b)\left[1 + \frac{1}{2}(a^\dagger a + b^\dagger b)\right],
\end{aligned} \tag{5.6}$$

i.e.,

$$S = \frac{1}{2}(a^\dagger a + b^\dagger b). \tag{5.7}$$

The relations (5.2) enlarge the on-site Hilbert space. To remain within the physical space, one has to satisfy (5.7).

A singlet ↑↓ in the sites (i, j) is given by

$$|\alpha\rangle_{ij} = \frac{1}{\sqrt{2}}\left(|\uparrow_i\downarrow_j\rangle - |\downarrow_i\uparrow_j\rangle\right), \tag{5.8}$$

and so

$$|\alpha\rangle_{ij} = \frac{1}{\sqrt{2}}\left(a_i^\dagger b_j^\dagger - b_i^\dagger a_j^\dagger\right)|0\rangle. \tag{5.9}$$

A general spin state $|S, m\rangle$ is generated by the relation

$$|S, m\rangle = \frac{(a^\dagger)^{S+m}(b^\dagger)^{S-m}}{\sqrt{(S+m)!}\sqrt{(S-m)!}}|0\rangle. \tag{5.10}$$

We can verify that

(a) If $S = ½$, $m = \pm ½$

$$|1/2, 1/2\rangle = a^\dagger|0\rangle, \qquad |1/2, -1/2\rangle = b^\dagger|0\rangle. \tag{5.11}$$

(b) If $S = 1$, $m = 1, 0, -1$

$$|1, 1\rangle = \frac{(a^\dagger)^2}{\sqrt{2}}|0\rangle, \qquad |1, -1\rangle = \frac{(b^\dagger)^2}{\sqrt{2}}|0\rangle, \quad |1, 0\rangle = a^\dagger b^\dagger|0\rangle. \tag{5.12}$$

(c) If $S = 2$, $m = 2, 1, 0, -1, -2$

$$\begin{aligned} |2, 2\rangle &= \frac{(a^\dagger)^4}{\sqrt{4!}}|0\rangle, \qquad |2, -2\rangle = \frac{(b^\dagger)^4}{\sqrt{4!}}|0\rangle, \\ |2, 1\rangle &= \frac{(a^\dagger)^3 b^\dagger}{\sqrt{3!}}|0\rangle \qquad |2, -1\rangle = \frac{a^\dagger(b^\dagger)^3}{\sqrt{3!}}|0\rangle, \\ |2, 0\rangle &= \frac{(a^\dagger)^2(b^\dagger)^2}{2}|0\rangle. \end{aligned} \tag{5.13}$$

Note that the condition (5.7) implies that the physical space is given by $n_a + n_b = 2S$, where $n_a = a^\dagger a$ and $n_b = b^\dagger b$. If a Schwinger boson, say a, condenses in some particular mode because it carries a spin index, such a condensate state spontaneously breaks the SU(2) symmetry. In this case, we can eliminate the a boson using the constraint (5.7)

$$b \leftrightarrow b, \qquad a \leftrightarrow \sqrt{2S - b^\dagger b}, \tag{5.14}$$

and we get the Holstein–Primakoff representation.

Let us define the bond operators

$$\hat{A}_{ij} \equiv \frac{1}{2}(a_i b_j - b_i a_j), \qquad \hat{B}_{ij} \equiv \frac{1}{2}\left(a_i^\dagger a_j + b_i^\dagger b_j\right). \tag{5.15}$$

We have then the identity

$$\vec{S}_i \cdot \vec{S}_j = :\hat{B}_{ij}^\dagger \hat{B}_{ij}: - \hat{A}_{ij}^\dagger \hat{A}_{ij}, \tag{5.16}$$

where :: denotes normal ordering. Both operators (5.15) are rotationally invariants. That is, they are invariant under any transformation SU(2) of the spinors $\begin{pmatrix} a_i \\ b_i \end{pmatrix}$. We can easily show that

$$\hat{B}_{ij}^{\dagger}\hat{B}_{ij} = 2(\vec{S}_i + \vec{S}_j)^2, \qquad \hat{A}_{ij}\hat{A}_{ij} = 2(\vec{S}_i - \vec{S}_j)^2, \tag{5.17}$$

revealing the character ferromagnetic and antiferromagnetic, respectively, of each term. Note that

$$\hat{A}_{ij} = -\hat{A}_{ji}. \tag{5.18}$$

$\hat{A}_{ij}^{\dagger}$ creates a spin-singlet in the oriented bond ij (it is called the valence bond (VB) order parameter, see section 5.8), and $\hat{B}_{ij}^{\dagger}$ implies a hopping of Schwinger bosons from site to site. $\hat{A}_{ij}^{\dagger}\hat{A}_{ij}$ is proportional to the number (0 or 1) of a singlet between sites i and j. The constraint (5.7) allows us to write

$$:\hat{B}_{ij}^{\dagger}\hat{B}_{ij}: + \hat{A}_{ij}^{\dagger}\hat{A}_{ij} = S^2. \tag{5.19}$$

So, we can eliminate one of the terms if it is of interest. The Hamiltonian

$$H = \pm J \sum_{\langle ij \rangle} \vec{S}_i \cdot \vec{S}_j, \tag{5.20}$$

can be written as

$$H = \pm J \sum_{\langle ij \rangle} \left(\hat{B}_{ij}^{\dagger}\hat{B}_{ij} - \hat{A}_{ij}^{\dagger}\hat{A}_{ij} \right) + \sum_i \lambda_i (a_i^{\dagger} a_i + b_i^{\dagger} b_i - 2S). \tag{5.21}$$

This Hamiltonian preserves the SU(2) spin-rotation invariance. The last term, where λ_i is a site-dependent Lagrange multiplier, restricts the number of bosons per site following the constraint (5.7). To proceed, we need to decouple the four operator terms in (5.21). By a judicious choice of decoupling, we can, in principle, describe any sort of ordered or disordered structure in the ground state of (5.21). The possibility of long-range magnetic order is incorporated in the theory by allowing the Schwinger bosons to condense.

5.2 Mean-field approximation

In the mean-field approximation (MFA) we take λ_i to be site-independent and use the decoupling

$$\hat{A}_{ij}^{\dagger}\hat{A}_{ij} \to \hat{A}_{ij}^{\dagger}\langle \hat{A}_{ij} \rangle + \langle \hat{A}_{ij}^{\dagger} \rangle \hat{A}_{ij} - |\langle \hat{A}_{ij}^{\dagger} \rangle|^2, \tag{5.22}$$

and a similar expression for the B term. The averages are taken in the mean-field state. In the MFA, we neglect bond operator fluctuations and replace $\langle \hat{A}_{ij} \rangle$ and $\langle \hat{B}_{ij} \rangle$ by complex bond parameters A_{ij} and B_{ij}. If we do not consider flux phases (see section 5.5 for the definition of flux), we can take the parameters to be real. The decoupling can also be done using the Hubbard–Stratonovich transformation [1]. The final result is the same.

A general Schwinger boson mean-field Hamiltonian with explicit global SU(2) symmetry must be of the form

$$H_{MF} = \sum_{\langle ij \rangle} J_{ij}\left(-A_{ij}^{*}\hat{A}_{ij} + B_{ij}^{*}\hat{B}_{ij} + h.c.\right) + \sum_{\langle ij \rangle} J_{ij}\left(|A_{ij}|^2 - |B_{ij}|^2\right) \\ + \lambda \sum_{i}\left(a_i^{\dagger}a_i + b_i^{\dagger}b_i - \kappa\right), \tag{5.23}$$

where the complex numbers $A_{ij} = -A_{ji}$, $B_{ij} = B_{ji}^{*}$ are the parameters of the mean-field ansatz (the set of parameters A_{ij}, B_{ij} appearing in the mean-field Hamiltonian is called an ansatz). The parameter κ can be continuously varied to interpolate between the classical limit ($\kappa = \infty$) and the extreme quantum limit ($\kappa \to 0$). Fixing $\kappa = 2S$ to study a spin S model is not necessarily the best choice. Note that A_{ij} is antisymmetric with respect to the interchange $i \to j$. So, for instance, in the Fourier transform along the x-direction of $a_i b_j$ we get (+) to the right and (−) to the left, which leads to: $e^{ik_x} - e^{-ik_y} = 2i \sin k_x$. We write

$$A_{ij} = \langle \hat{A}_{ij} \rangle, \qquad B_{ij} = \langle \hat{B}_{ij} \rangle. \tag{5.24}$$

At the mean field, the local constraint (5.7) is enforced only on the average via the Lagrange multiplier λ. Note that the present formalism does not impose any preferred spin direction: giving a finite expectation value $A_{ij} \neq 0$ to the operator $\hat{A}_{ij}$ does not break the SU(2) symmetry.

One of the most used approaches is searching for an optimal A_{ij} and B_{ij} self-consistently solving for stationary points of the mean-field free energy. The mean-field energy of a gas of bosons is given by

$$F_{MF} = N\beta^{-1}\sum_{k} \ln\left(1 - e^{-\beta\omega_k}\right) + E_g, \tag{5.25}$$

where E_g is the ground-state energy of the Hamiltonian (5.23). The conditions (5.24) should be obeyed to enforce self-consistency, which is equivalent to

$$\frac{\partial F_{MF}}{\partial A_{ij}} = 0, \text{ and } \qquad \frac{\partial F_{MF}}{\partial B_{ij}} = 0,$$

together with the condition $\langle \hat{n}_i \rangle = \kappa \leftrightarrow \frac{\partial F_{MF}}{\partial \lambda} = 0$.

As is often the case with mean-field theory, the self-consistent equations possess several solutions. To compare the various solutions, one can, at a first approximation, compare the mean-field energies (but note that these energies are not variational).

As was said before, we can use $\hat{A}_{ij}$ or $\hat{B}_{ij}$, and keeping only $\hat{A}_{ij}$ is a widespread practice in the literature. However, Mezio *et al* [2] have shown that using only $\hat{A}_{ij}$ gave incorrect results to the triangular lattice. Even on the square lattice, the simultaneous use of both operators improved the result for the ground-state energy [3]. In frustrated systems where disordered phases could appear, the use of both operators is more convenient [4, 5].

Auerbach [1] extended the SU(2) Schwinger boson representation to SU($\mathcal{N}$) with $\mathcal{N}$ large by increasing the number of Schwinger boson flavors from 2 to $\mathcal{N}$. The bond operator is generalized to: $\hat{A}_{ij} = \sum_{\alpha=1}^{N} a_{i\alpha} a_{j\alpha}$. The mean-field theory is strictly valid only in the $\mathcal{N} \to \infty$ limit, as we show below. We have the exact result

$$\hat{A}_{ij}^{\dagger}\hat{A}_{ij} = \hat{A}_{ij}^{\dagger}\langle\hat{A}_{ij}\rangle + \left\langle\hat{A}_{ij}^{\dagger}\right\rangle\hat{A}_{ij} - \left\langle\hat{A}_{ij}^{\dagger}\right\rangle\langle\hat{A}_{ij}\rangle + \left(\hat{A}_{ij}^{\dagger} - \left\langle\hat{A}_{ij}^{\dagger}\right\rangle\right)(\hat{A}_{ij} - \langle\hat{A}_{ij}\rangle).$$

All terms except the last one are of order $\mathcal{N}^2$. By the central limit theorem, the last term is a product of terms of order square root of $\mathcal{N}$, and thus, one order less than the others. In the large $\mathcal{N}$ limit, it can be neglected, and we get the mean-field result (5.22). However, the mean-field theory can be used as an analytic approach even for $S = ½$, since it captures various disordered and exotic ground states. The large $\mathcal{N}$ approach handles strong local interactions in terms of constraints, and at the mean-field level, the constraints are enforced only on average. Higher-order corrections systematically reintroduce their effects in $1/\mathcal{N}$, including the disordering effects of quantum fluctuations. We can treat the last term in the exact expansion in perturbation theory, and the $1/\mathcal{N}$ corrections to the mean-field theory can be calculated using Feynman diagrams, which describe interactions between the free quasiparticles of the mean-field theory. Finding a large-$\mathcal{N}$ ground state amounts to solving a classical minimization problem. However, it is rather difficult to include the effect of fluctuations beyond the mean-field approximation. Although we will not consider the implications of $1/\mathcal{N}$ corrections, they present a way to extend the mean-field results significantly when the mean-field theory predicts pathological behavior.

The SU($\mathcal{N}$) approach proposed by Auerbach [1] is valid only for ferromagnets. Bipartite antiferromagnets can also be studied in SU($\mathcal{N}$) by performing a special transformation on one sublattice. Read and Sachdev [6] extended the theory by duplicating $\mathcal{N}$ times the two species of bosons at each site to the group Sp($\mathcal{N}$), called the symplectic group, which is a subgroup of SU($\mathcal{N}$). In the group Sp($\mathcal{N}$) the elements reverse sign under time reversal, just like the SU(2) spins. The generalized bond operator is now written as $\hat{A}_{ij} = \frac{1}{2}\sum_{\alpha=1}^{N}(a_{i\alpha}b_{j\alpha} - b_{i\alpha}a_{j\alpha})$. For a definition of the group Sp($\mathcal{N}$) see appendix A. For $\mathcal{N} = 1$, the theory becomes the Heisenberg model with SU(2) symmetry. I will not use the large $\mathcal{N}$ theory in this book, and the presentation above was to give some information to the reader. For further details, one should consult the references.

5.3 Ferromagnet

This case is trivial. We use the operator $\hat{B}$. From (5.16) and (5.19) we have

$$H = -J\sum_{\langle ij\rangle}\vec{S}_i \cdot \vec{S}_j = -J\sum_{\langle ij\rangle}\left(2\colon \hat{B}_{ij}^{\dagger}\colon B_{ij} - S^2\right) \tag{5.26}$$

$$H_{MF} = -2J\sum_{\langle ij\rangle}\left[\frac{B}{2}\left(a_ia_j^\dagger + a_i^\dagger a_j + b_ib_j^\dagger + b_i^\dagger b_j\right) - B^2\right] + \lambda\sum_i(a_i^\dagger a_i + b_i^\dagger b_i - 2S). \tag{5.27}$$

Or

$$H_{MF} = -JB\sum_{\langle ij\rangle}\left(a_ia_j^\dagger + a_i^\dagger a_j + b_ib_j^\dagger + b_i^\dagger b_j\right) + \lambda\sum_i\left(a_i^\dagger a_i + b_i^\dagger b_i\right) + NzJB^2 - 2N\lambda S, \tag{5.28}$$

where N is the number of sites. Fourier transforming (5.28) we get

$$H_{MF} = \sum_k \omega_k(a_k^\dagger a_k + b_k^\dagger b_k) + NzB^2 - 2N\lambda S, \tag{5.29}$$

where

$$\omega_k = \lambda - zJB\gamma_k. \tag{5.30}$$

Equation (5.25) is here written as

$$F_{MF} = N\beta^{-1}\sum_k \ln(1 - e^{-\beta\omega_k}) + NzJB^2 - 2N\lambda S. \tag{5.31}$$

Minimizing (5.31) with respect to λ and B we obtain:

$$\frac{1}{N}\sum_k n_k = S, \tag{5.32}$$

$$\frac{1}{N}\sum_k n_k\gamma_k = B. \tag{5.33}$$

In the thermodynamic limit, we take $N \to \infty$, and integrals can substitute the sums. As we take $T \to 0$, the Bose function vanishes for all $\vec{k} \neq 0$. There must be a macroscopic occupation at $\vec{k} = 0$ to satisfy the constraint (5.32):

$$S \approx \frac{1}{N}n_0. \tag{5.34}$$

That is, there is a Bose condensation at $T = 0$ at $\vec{k} = 0$. The Bose condensation is complete because all available Schwinger boson density ($2S$ per site) is accumulated in that single mode. Auerbach [1] showed that in the limit $T \to 0$ the dispersion relation ω_k reduces to the one for the ferromagnetic spin waves: $\omega_k = zJS(1 - \gamma_k)$.

Note that the boson condensation occurs only when the number of bosons is conserved, as is the case with Schwinger bosons. Since they cannot disappear, they condense in the ground state at $T = 0$. That does not happen with magnons of the

Holstein–Primakoff representation since there is no conservation rule for them. If there is a gap in the dispersion relation, equation (5.32) can be satisfied without the necessity of condensation. As we saw before, if there is a gap in the Heisenberg model's excitation spectrum, the ground state must be disordered.

Equation (5.30) can be written as

$$\omega_k = zB\left(1 - \gamma_k + \frac{1}{4z}\kappa^2\right), \qquad \kappa^2 = 4z\left(\frac{\lambda}{zB} - 1\right). \tag{5.35}$$

In general, we must solve the self-consistent equations numerically. For the Heisenberg ferromagnet and antiferromagnet in one and two dimensions, Auerbach [1] obtained some analytical expressions, and I will not present his result here. However, I would like to address just one point: the calculation of the correlation length. The standard spin-wave theories can be used only in the ordered phase and therefore give infinite correlation length (although the correlation length can be estimated for the disordered phase with some suppositions). We can get analytical expressions for the correlation length in disordered phases in just a few cases. Since the Schwinger boson formalism can be used in disordered phases, it is a convenient tool for this kind of calculation. Here, I will follow Auerbach [1] to calculate the ferromagnet's correlation length in two dimensions. He has shown that in this case, one has

$$\kappa \approx \sqrt{16\pi}\,\exp\left(-\frac{2\pi JS^2}{T}\right), \qquad B = JS + O(T/JS). \tag{5.36}$$

The free boson expression gives the mean-field spin correlation function

$$\left\langle S_i^+ S_j^- \right\rangle = \left\langle a_i^\dagger b_i b_j^\dagger a_j \right\rangle = |R_{ij}|^2 + S\delta_{ij}, \tag{5.37}$$

where

$$R_{ij} = \frac{1}{N}\sum_k n_k e^{i\vec{k}\cdot(\vec{r}_i - \vec{r}_j)}. \tag{5.38}$$

For small values of $\vec{k}$ we have $n_k \approx T/\omega_k$, $\gamma_k \approx 1 - k^2/4$ and so we can write (5.38) as

$$R_{ij} \approx \frac{T}{JS}\int \frac{d^2\vec{k}}{(2\pi)^2}\frac{e^{i\vec{k}\cdot(\vec{r}_i - \vec{r}_j)}}{\kappa^2/4 + k^2}. \tag{5.39}$$

At large distances, we get

$$R_{ij} \propto \left(|\vec{r}_i - \vec{r}_j|/\xi\right)^{-1/2} \exp\left(-|\vec{r}_i - \vec{r}_j|\kappa/2\right)\left(1 + \frac{2}{|\vec{r}_i - \vec{r}_j|\kappa} + \cdots\right). \tag{5.40}$$

Inserting (5.40) into (5.37) we find

$$\left\langle S_i^+ S_j^- \right\rangle \propto \frac{\xi}{|\vec{r}_i - \vec{r}_j|} e^{-|\vec{r}_i - \vec{r}_j|/\xi}, \tag{5.41}$$

where $\xi = \kappa^{-1}$ is the spin correlation length. We see that, in agreement with the Mermin–Wagner theorem, there is no long-range order at $T > 0$.

5.4 Antiferromagnet

There is more than one way to treat the problem. First, I will follow Auerbach [1] and consider the antiferromagnet on a bipartite lattice with sublattices A and B and nearest-neighbor interactions and use only the operator $\hat{A}_{ij}$. As I said before, this operator is antisymmetric to the interchange of $i \to j$. As we know, one has a Néel order in the ground state (at least in 2d and 3d). So, we apply a spin rotation by π about the y-axis on sublattice B, which changes

$$a_j \to -b_j, \qquad b_j \to a_j. \tag{5.42}$$

This transformation preserves the constraint (5.7). The bond operator transforms into a symmetric operator

$$A_{ij} \to A_{ij} = \frac{a_i a_j + b_i b_j}{2}. \tag{5.43}$$

The Hamiltonian is then written as

$$H = -J \sum_{\langle ij \rangle} \left(2 \hat{A}_{ij}^\dagger \hat{A}_{ij} - S^2 \right). \tag{5.44}$$

The mean-field Hamiltonian becomes

$$\begin{aligned} H_{MF} = JA \sum_{\langle ij \rangle} \left(a_i^\dagger a_j^\dagger + a_i a_j + b_i^\dagger b_j^\dagger + b_i b_j \right) \\ + \lambda \sum_i (a_i^\dagger a_i + b_i^\dagger b_i) + NzJA^2 - 2NS\lambda. \end{aligned} \tag{5.45}$$

Fourier transforming, we get

$$\begin{aligned} H_{MF} = {} & \sum_k \lambda \left(a_k^\dagger a_k + b_k^\dagger b_k \right) \\ & + \frac{JzA}{2} \sum_k \gamma_k \left(a_k^\dagger a_{-k}^\dagger + a_k a_{-k} + b_k^\dagger b_{-k}^\dagger + b_k b_{-k} \right) \\ & + NJzA^2 - 2NS\lambda. \end{aligned} \tag{5.46}$$

Using the Bogoliubov transformation

$$a_k = \cosh\theta_k \alpha_k + \sinh\theta_k \alpha_{-k}^\dagger, \qquad b_k = \cosh\theta_k \beta_k + \sinh\theta_k \beta_{-k}^\dagger, \tag{5.47}$$

we write (5.46) as

$$
\begin{aligned}
H_{MF} &= \frac{1}{2}\sum_k (\lambda \cosh 2\theta_k + zJA\gamma_k \sinh 2\theta_k)\left(\alpha_k^\dagger \alpha_k + \alpha_k \alpha_k^\dagger + \beta_k^\dagger \beta_k + \beta_k \beta_k^\dagger\right) \\
&\quad (\lambda \sinh 2\theta_k + zJA\gamma_k \cosh 2\theta_k)\left(\alpha_k^\dagger \alpha_{-k}^\dagger + \alpha_k \alpha_{-k} + \beta_k^\dagger \beta_{-k}^\dagger + \beta_k \beta_{-k}\right) \\
&\quad + NzJA^2 - 2N\left(S + \frac{1}{2}\right)\lambda.
\end{aligned}
\tag{5.48}
$$

The condition that the anomalous term should vanish leads to

$$\tanh 2\theta_k = -\frac{zJA\gamma_k}{\lambda}. \tag{5.49}$$

The final result is

$$H_{MF} = \sum_k \omega_k \left(\alpha_k^\dagger \alpha_k + \beta_k^\dagger \beta_k + \frac{1}{2}\right) + NzJA^2 - 2N\left(S + \frac{1}{2}\right)\lambda, \tag{5.50}$$

where

$$\omega_k = \sqrt{\lambda^2 - (zJA\gamma_k)^2}\,. \tag{5.51}$$

Using $\ln(1 - e^{-x}) = \ln e^{-x/2}(e^{x/2} - e^{-x/2}) = -x/2 + \ln 2 \sinh(x/2)$, we can write the mean-field energy as

$$F_{MF} = \beta^{-1}\sum_k 2 \ln\left[2 \sinh\left(\frac{\beta\omega_k}{2}\right)\right] - 2N\left(S + \frac{1}{2}\right)\lambda + NzJA^2. \tag{5.52}$$

Minimizing (5.52) with relation to λ and A gives

$$\frac{1}{N}\sum_k \frac{\lambda}{\omega_k}\left(n_k + \frac{1}{2}\right) = S + \frac{1}{2}, \tag{5.53}$$

$$\frac{1}{N}\sum_k \frac{z^2 JA\gamma_k^2}{\omega_k}\left(n_k + \frac{1}{2}\right) = zA. \tag{5.54}$$

The excitations in the Schwinger boson approach are not the conventional magnons of the spin-wave theory. The Schwinger boson Hamiltonian is quartic in Bose operators so that magnon–magnon interactions are incorporated into the theory already at the mean-field order. The problem was well discussed by Auerbach [1]. He shows that the Schwinger boson mean-field ground state for the nearest-neighbor model is disordered in one dimension and has long-range order in two and three dimensions. The Schwinger boson mean-field theory fails for half-odd integer spin chains. As mentioned before, the one-dimensional antiferromagnet is a complex problem.

Yoshioka [7] treated the same problem using only the VB order parameter $\hat{A}_{ij}$ without applying the spin rotation at sublattice B (he did not use a bipartite lattice). He commented that it is not strange that the Néel order appears in the ground state, since the VB state can, in general, express any spin-singlet state, and the Néel order is such a state. When Auerbach divided the lattice into two sublattices, there was an implicit assumption of the Néel order's existence. For the case of the square lattice, this assumption is reasonable and justified. For other lattices, the nature of the long-range order (if it exists) is not clear.

Now I will use the two operators to describe the same model. For a pedagogical purpose, I will treat the nearest-neighbor problem at zero temperature. However, the procedure is more useful when we have, for instance, second and even third nearest-neighbor spins. Using both operators, we write the Hamiltonian, after the mean-field decoupling, as

$$\begin{aligned} H = J\sum_{\langle ij\rangle}\Big[\left(B_{ij}^*\hat{B}_{ij} - A_{ij}^*\hat{A}_{ij} + h.c.\right)+\left(-B_{ij}^*B_{ij} + A_{ij}^*A_{ij} + h.c.\right)\Big] \\ +\lambda\sum_i(a_i^\dagger a_i + b_i^\dagger b_i - 2S). \end{aligned} \tag{5.55}$$

After a Fourier transformation, we get

$$\begin{aligned} H_{MF} = \sum_k\big[B(k) + \lambda\big]\left(a_k^\dagger a_k + b_k^\dagger b_k\right) \\ - i\sum_k A(k)\left(a_k b_{-k} + a_k^\dagger b_{-k}^\dagger\right) - 2\lambda NS + 2N(A^2 - B^2), \end{aligned} \tag{5.56}$$

where

$$A(\vec{k}) = \frac{J}{2}\sum_\delta \sin(\vec{k}\cdot\vec{\delta})A_\delta, \qquad B(\vec{k}) = \frac{J}{2}\sum_\delta \cos(\vec{k}\cdot\vec{\delta})B_\delta, \tag{5.57}$$

and the sums are performed over the $\vec{\delta}$ vectors connecting the pair of sites coupled by J.

We have: $A_{-\vec{\delta}} = -A_{\vec{\delta}}$, $B_{-\vec{\delta}} = B_{\vec{\delta}}$. After a Bogoliubov transformation

$$\alpha_k^\dagger = \cosh\theta_k a_k^\dagger - \sinh\theta_k b_{-k}, \qquad \beta_k^\dagger = \cosh\theta_k b_k^\dagger - \sinh\theta_k a_{-k}, \tag{5.58}$$

with

$$\tanh\theta_k = -\frac{A(\vec{k})}{B(\vec{k}) + \lambda}, \tag{5.59}$$

we get

$$H_{MF} = \sum_k \omega_k\left(\alpha_k^\dagger\alpha_k + \beta_k^\dagger\beta_k + 1\right) - N\lambda(1 + 2S), \tag{5.60}$$

where

$$\omega_k = \sqrt{[B(\vec{k}) + \lambda]^2 - A^2(\vec{k})}. \tag{5.61}$$

Minimizing F_{MF} at zero temperature, we find

$$\frac{1}{2N}\sum_k \frac{A(\vec{k})}{\omega_k}(1 + 2n_k)\sin(\vec{k}\cdot\vec{\delta}) = A_{\vec{\delta}}, \tag{5.62}$$

$$\frac{1}{2N}\sum_k \frac{B(\vec{k}) + \lambda}{\omega_k}(1 + 2n_k)\cos(\vec{k}\cdot\vec{\delta}) = B_{\vec{\delta}}, \tag{5.63}$$

$$\frac{1}{2N}\sum_k \frac{B(\vec{k}) + \lambda}{\omega_k}(1 + 2n_k) = \frac{1}{2} + S. \tag{5.64}$$

These equations were solved numerically by Ceccato *et al* [3]. They found $B = 0$ and the ground-state energy $E_g = -0.335$ in agreement with quantum Monte Carlo calculation. The value using the Auerbach [1] approach is $E_g = -0.420$.

5.5 Gauge transformation

In this section, I will indicate the Schwinger bosons by b_α, with $b_1 = a$, $b_2 = b$. Let us consider the local U(1) gauge transformation $b_{n\alpha} \rightarrow e^{i\theta_n}b_{n\alpha}$. The Hamiltonian remains invariant if we perform the transformation

$$A_{ij} \rightarrow e^{i(\theta_i+\theta_j)}A_{ij}, \qquad B_{ij} \rightarrow e^{i(-\theta_i+\theta_j)_i}B_{ij}. \tag{5.65}$$

If a gauge transformation connects two mean-field ansatzes after a symmetry operation, the two ansatzes will describe the same physical state and should be considered identical. The set of all transformations that leave a mean-field ansatz invariant is called the projective symmetry group (PSG). Some elements of the PSG are a pure local transformation of the kind (5.65). The set of such elements form a subgroup of the PSG called the invariant gauge group (IGG). If A_{ij} and B_{ij} are both non-zero, the IGG group has elements: identity and $b_{i\alpha} \rightarrow -b_{i\alpha}$. That is the Z_2 group. If the B_{ij} are non-zero while all A_{ij} vanish, the IGG will be a U(1) group $b_{i\alpha} \rightarrow e^{i\varphi}b_{i\alpha}$ where φ is a site-independent constant. The sum of the phases of an operator around a plaquette is called *flux*. If an operator is written as $\chi_{ij} = \chi_0 e^{i\theta_{ij}}$, the flux is defined as $\theta_{ij} + \theta_{jk} + \theta_{kl} + \theta_{li}$ in the plaquette *ijkl*.

5.6 Frustration

Before continuing with the Schwinger bosons, I will briefly discuss the concept of frustration, which will be useful in the following. An excellent reference in the subject is Schmidt and Thalmeier [8]. There are two types of spin-exchange frustration. First, one has geometric frustration when only antiferromagnetic nearest-neighbor (n. n.) bonds are present, and the exchange bonds cannot be simultaneously minimized because of local geometric constraints, enforced by the lattice coordination as, e. g., in the triangular or kagome lattice. The other case is the frustration due to competition when the n. n. interaction themselves are unfrustrated, as in the square lattice. Still, the exchange bonds to the next nearest (n. n. n.) or further neighbors introduce a conflict of spin orientation. In the quantum Heisenberg model, one has the simultaneous influence of frustration and quantum fluctuations (zero-point motion of spin waves) at zero temperature complemented by thermal fluctuations (thermally excited spin-wave modes) at finite temperature. Frustration and competition between interactions is a highly debated issue in the field of quantum magnetism. A good review paper on this topic is Misguich [9].

Suppose the ground state of a classical Heisenberg model has no long-range magnetic order because of frustration. In that case, it does not necessarily imply that this remains true at $T > 0$ or for the quantum version of the model [10]. The spectrum of fluctuations generally depends on the ground state. That can lead to a selection mechanism known as *order by disorder* that can lead to long-range magnetic order.

One model of great interest is the two-dimensional square lattice frustrated spin-1/2 Heisenberg antiferromagnet, commonly referred to as the J_1–J_2 model. This model has been studied extensively by various analytical and numerical techniques [9]. For $J_2 = 0$, the ground state is antiferromagnetically ordered at zero temperatures as was said before. The next nearest-neighbor interaction J_2 (along the diagonals) induces frustration and breaks the antiferromagnetic order.

In the classical or mean-field picture of the magnetic order, the spin structure is uniquely determined by minimizing the mean-field ground-state energy $E_{cl} = NS^2J(\vec{Q})$ where $\vec{Q}$ is the modulation vector of the magnetic phase and $J(\vec{k})$ is given by $J(\vec{k}) = J_1(\cos k_x + \cos k_y) + 2J_2 \cos k_x \cos k_y$.

Minimizing E_{cl} with relation to $\vec{Q}$ leads to three possible magnetic phases: Ferromagnet (FM): $\vec{Q} = (0, 0)$, Néel antiferromagnet (NAF) $\vec{Q} = (\pi, \pi)$ and collinear antiferromagnet (CAF) $\vec{Q} = (\pi, 0)$ or $\vec{Q} = (0, \pi)$, where the neighboring spins align ferromagnetically along one axis of the square lattice and antiferromagnetically along with the other (see table 5.1). In fact, the classical ground state is infinitely degenerate for $J_2/J_1 > 1/2$, and the energy is independent of the angle between the spins on the two sublattices. Quantum fluctuations lead to the collinear

Table 5.1. The condition of existence of the three magnetic phases

Phase	Condition
Ferromagnet	$J_1 \leqslant 0,\ J_2/\lvert J_1\rvert \leqslant 1/2$
Antiferromagnet	$J_1 \geqslant 0,\ J_2/\lvert J_1\rvert \leqslant 1/2$
Collinear AF	$J_2 \geqslant 0,\ J_2/\lvert J_1\rvert \geqslant 1/2$

ground state. At zero temperature, the spin orientations of the three types of classical ground-state configurations exhibited by the model are shown in figure 5.1.

The interplay of quantum fluctuations and frustration will strongly modify the classical picture. It has been found that for small $\eta = J_2/J_1$, the ground state is still Néel ordered, but it undergoes a second-order quantum phase transition at a certain $\eta = \eta_{1C}$ from the Néel to a magnetically disordered state. For η larger than another critical value η_{2C}, one has a collinear ordered state. There is substantial evidence that the ground state of the quantum disordered phase has no long-range magnetic order for $0.4 < \eta < 0.6$ for $S = ½$. There is, however, a good deal of disagreement over the exact positions of the critical points, and the nature of the magnetically disordered region is still under debate.

It happens that the simple form of the J_1–J_2 model is rarely fully adequate to treat realistic systems. In particular, lattice distortions may lead to further anisotropies of the exchange interactions. The spin–orbit interactions can also cause anisotropies of the symmetric exchange [8].

As was mentioned before, the spin-wave formalism cannot be used to study magnetically disordered phases. Still, it can be applied to the ordered phases of frustrated systems, and it can indicate the transition to the disordered ones.

In the study of frustrated magnets, it is usual to plot χ^{-1} as a function of temperature, which at high temperature has the linear form $\chi^{-1} \propto (T - \theta_{CW})$, where the Curie–Weiss constant θ_{CW} indicates the magnet's sign and strength of interactions. θ_{CW} is negative in an antiferromagnet. In the absence of frustration, magnetic order appears below the Néel temperature $T_N \sim |\theta_{CW}|$. Geometrically frustrated magnets do not order at T_N. Instead, they remain in the paramagnetic phase to a much lower temperature at $T_C \ll |\theta_{CW}|$ (a large value for $f \equiv |\theta_{CW}|/T$ is a signature of frustration). These magnets are strongly correlated in the region $T_C \ll T \ll T_N$. However, long-range order is absent. A remarkable characteristic of frustrated magnets is the large accidental degeneracy of ground states, which manifests as zero modes that are not associated with any symmetry breaking (distinct from the Goldstones modes). In most cases, it is not easy to treat these zero modes analytically since spin configurations in the classical ground states do not have long-range order, and we should resort to numerical calculations.

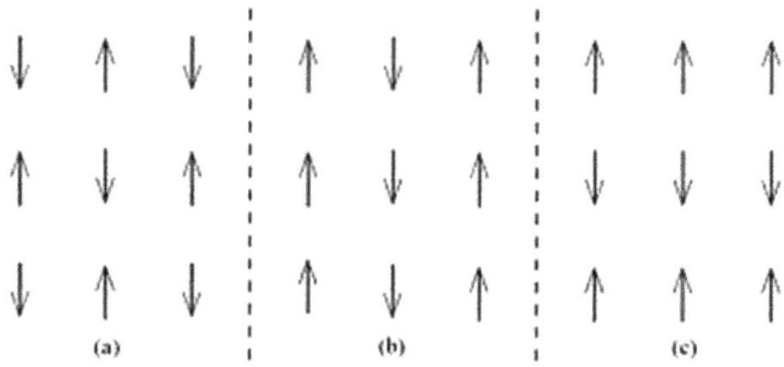

Figure 5.1. Classical ground states for the model studied in this paper: (a) Néel (π, π), (b) collinear (π, 0), (c) collinear (0, π).

5.7 Schwinger boson and the J_1–J_2 model

Following Yang and Wang [11], I will study the square lattice J_1–J_2 antiferromagnetic Heisenberg model, mentioned in the last section, using the Schwinger boson formalism. The Hamiltonian is

$$H = J_1 \sum_{\langle ij \rangle} \vec{S}_i \cdot \vec{S}_j + J_2 \sum_{\langle\langle ij \rangle\rangle} \vec{S}_i \cdot \vec{S}_j. \tag{5.66}$$

As was mentioned in the last section, the ground state of this model is the Néel state for $J_2/J_1 \leqslant 0.4$ and shows a collinear order for $J_2/J_1 \geqslant 0.6$. It is not the goal of this section to study the disordered region, but only to present an example of the use of the Schwinger bosons. Several mean-field ansatzes are treated in the literature, and here I will show only one for pedagogical purposes.

As before, we will write the mean-field Hamiltonian as

$$H_{MF} = \sum_{ij} J_{ij}\left(-A_{ij}^* \hat{A}_{ij} + B_{ij}^* \hat{B}_{ij} + h \cdot c\right) + \sum_{ij} J_{ij}\left(|A_{ij}|^2 - |B_{ij}|^2\right) - \lambda \sum_i (\hat{n}_i - \kappa), \tag{5.67}$$

where $\langle \hat{n}_i \rangle \equiv \kappa = 2S$. We use an ansatz where A_1 and B_2 are real, and B_1 is pure imaginary. The ansatz is given by

$$\begin{aligned} A_{(x,\,y)(x+1,\,y)} &= -A_{(x,\,y)(x,\,y+1)} = A_1, \qquad B_{(x,\,y)(x+1,\,y)} = -B_{(x,\,y)(x,\,y+1)} = B_1 \\ B_{(x,\,y)(x+1,\,y+1)} &= -B_{(x,\,y)(x-1,\,y+1)} = B_2. \end{aligned} \tag{5.68}$$

After a Fourier transformation, H_{MF} is written as

$$H_{MF} = \sum_k \psi_k^\dagger D_k \psi_k + N\left[\lambda(1 + 2S) + 2J_1\left(|A_1|^2 - |B_1|^2\right) - 2J_2\,|B_2|^2\right], \tag{5.69}$$

where $\psi_k = (a_k, b_{-k}^\dagger)^T$ and

$$D_k = (-2J_2 B_2 f_1 - \lambda)I - J_1 A_1 f \sigma_y - \mathrm{Im}(J_1 B_1) f \sigma_z, \tag{5.70}$$

where

$$f_1 = \sin k_x \sin k_y, \qquad f = \sin k_x - \sin k_y. \tag{5.71}$$

Performing a Bogoliubov transformation, we get

$$H_{MF} = \sum_k \omega_k\left(\alpha_k^\dagger \alpha_k + \beta_k^\dagger \beta_k + 1\right) + N[\lambda(1 + \kappa) + 2J_1(|A_1|^2 - |B_1|^2) - 2J_2\,|B_2|^2], \tag{5.72}$$

where

$$\omega_{k\pm} = \sqrt{(2J_2 B_2 f_1 + \lambda)^2 - (J_1 A_1 f)^2} \pm J_1 \mathrm{Im}(B_1 f). \tag{5.73}$$

The self-consistent equations at $T = 0$ are

$$\begin{aligned}
1 + \kappa &= -\int_{BZ} \frac{1}{2}\left(\frac{\partial \omega_{k+}}{\partial \lambda} + \frac{\partial \omega_{k-}}{\partial \lambda}\right) d^2k, \\
4A_1 &= -\int_{BZ} \frac{1}{2}\left(\frac{\partial \omega_{k+}}{\partial A_1} + \frac{\partial \omega_{k-}}{\partial A_1}\right) d^2k, \\
4|B_1| &= \int_{BZ} \frac{1}{2}\left(\frac{\partial \omega_{k+}}{\partial |B_1|} + \frac{\partial \omega_{k-}}{\partial |B_1|}\right) d^2k, \\
4B_2 &= -\int_{BZ} \frac{1}{2}\left(\frac{\partial \omega_{k+}}{\partial B_2} + \frac{\partial \omega_{k-}}{\partial B_2}\right) d^2k.
\end{aligned} \tag{5.74}$$

Yang and Wang [11], using different ansatzes, found six spin-liquid states that could be relevant to the J_1–J_2 Heisenberg model (for more details, the reader should consult their paper).

5.8 Valence bonds

In some magnetic models with frustrated interactions, low spin quantum numbers, and especially in small dimensions, long-range magnetic order (defined as we saw before as the existence of a non-zero on-site magnetization) is not the most general situation. Although we cannot find the exact quantum ground state of the antiferromagnetic Heisenberg model in many cases, it is sometimes possible to find a good approximation from a suitable variational ansatz when the spin-wave formalism does not work very well. Generalizing the exact result of the four site spin antiferromagnetic Heisenberg model discussed in chapter 1, we might expect that the ground state is given by a superposition of many possible singlet configurations. Many such systems can be represented in terms of spin paired into rotationally invariant singlets, called *valence bonds* (VB) [1, 12–14]. The VB on a pair of sites i and j is given by

$$|(ij)\rangle \equiv \left(a_i^\dagger b_j^\dagger - b_i^\dagger a_j^\dagger\right)|0\rangle, \tag{5.75}$$

where a_i and b_i are Schwinger bosons on site i and $|0\rangle$ is the vacuum state without any bosons. This state minimizes the energy of the antiferromagnetic coupling between the two spins

$$\langle (ij)| \vec{S}_i \cdot \vec{S}_j \,|(ij)\rangle = -3/4. \tag{5.76}$$

Note that we can still have long-range order at $T = 0$. Two neighboring spins first pair themselves in a singlet, forming a valence bond (which we call dimers). These dimers can then order among themselves, creating a valence bond solid (VBS), which breaks translational symmetry. As in the Néel order, this phase can be described by a local order parameter (figure 5.2).

If we assign a VB to each pair of a given partition of the set of sites of an antiferromagnetic lattice, we can define a variational wave function as

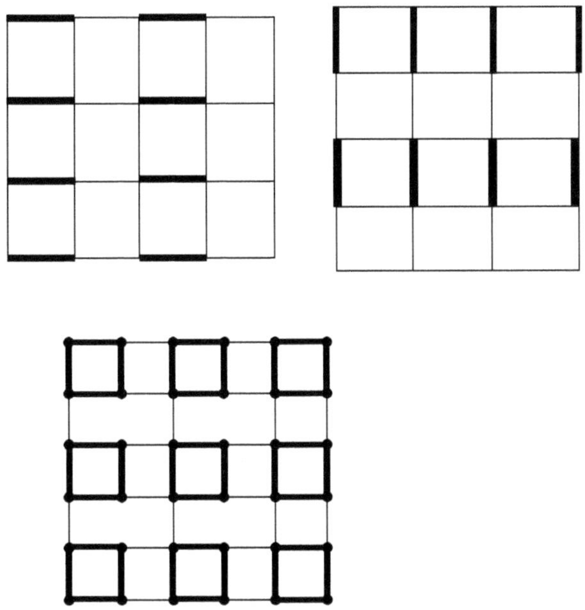

Figure 5.2. A VBS breaks lattice symmetry.

$$|VB\rangle = \sum_{\alpha} c_{\alpha} |\alpha\rangle, \tag{5.77}$$

where c_α are variational parameters,

$$|\alpha\rangle = \prod_{(ij)\in\Lambda_\alpha} |(ij)\rangle, \tag{5.78}$$

and Λ_α is a particular configuration of bonds (ij) on the lattice such that exactly $2S$ bonds end at each lattice site. One sees that every term in $|\alpha\rangle$ contains exactly $2S$ creation operators at each site, i.e. $|\alpha\rangle$ satisfies the constraint $a_i^\dagger a_i + b_i^\dagger b_i = 2S$ for local spin quantum number S and is thus (generally not normalized) an appropriate state of the Heisenberg model with spin S. Like $|\alpha\rangle$, the superposition (5.77) is a spin-singlet.

Note that although $|VB\rangle$ is a singlet, the nature of the spin correlations depends on the weight c_α. In particular, if this state has significant weight for a configuration Λ_α, the spin correlation can even be long-ranged ordered (Néel ordered). If the weight can be neglected when the valence bond exceeds a finite $\xi \sim O(1)$ length, the spin correlations are expected to decay exponentially.

Even if we allow only nearest-neighbor valence bonds, many possible dimers covering the entire lattice are possible. For each dimer covering of the lattice, there is a valence bond basis state. The Hamiltonian flips a pair of vertical bonds on a plaquette to a pair of horizontal ones and vice-versa. If this is the case, the resulting ground state would not have a long-range order. Such states were called *resonance valence bond* (RVB) states by Anderson (figure 5.3).

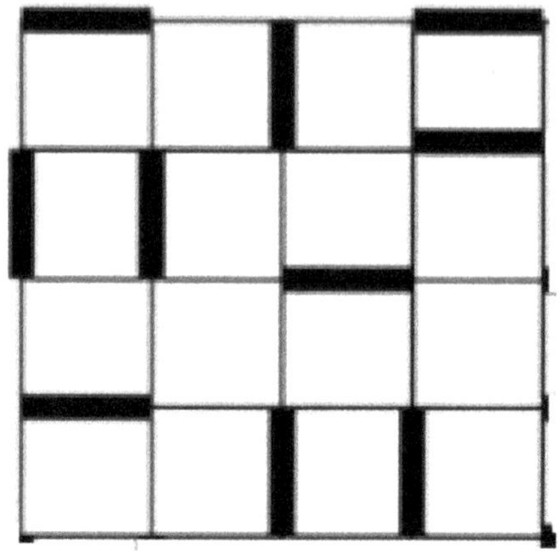

Figure 5.3. The RVB state is a superposition of different partitioning of spins into valence bond.

In a paramagnetic phase, a distinction should be made between spin systems that possess long-range order (LRO) in the VB arrangements (the valence bond crystals mentioned above) and spin systems that do not develop long-range order in any local order parameter at any temperature. The latter systems are called *spin liquids.* Such systems have no LRO in spin, dimer, or any (finite) higher-order correlation functions. The spinons excitations are deconfined. In inelastic neutron scattering experiments, spinons are created in pairs because a neutron changes S^z_{tot} by ± 1. The two spinons of a given pair can then propagate independently. We will return to spin liquids in chapter 10.

Let us consider the case of spin ½, on a bipartite valence bond configuration Λ_α, i.e., $(ij) \in \Lambda_\alpha$ only if $i \in A$ and $j \in B$ or $i \in B$ and $j \in A$ for sublattices A and B. Without loss of generality, we assume $i \in A$ and $j \in B$.

The state $|\alpha\rangle$ can then be written as

$$|\alpha\rangle = \prod_{(ij)\in\Lambda_\alpha} \frac{|\uparrow\rangle_i|\downarrow\rangle_j - |\downarrow\rangle_i|\uparrow\rangle_j}{\sqrt{2}}. \tag{5.79}$$

The spin correlation function in this state is given by

$$\langle\alpha|\vec{S}_k \cdot \vec{S}_l|\alpha\rangle - \langle\alpha|\vec{S}_k|\alpha\rangle.\ \langle\alpha|\vec{S}_l|\alpha\rangle. \tag{5.80}$$

For $k = l$ we have $\langle\alpha|\vec{S}_k \cdot \vec{S}_l|\alpha\rangle = S(S+1) = 3/4$. In the following, we assume $k \neq l$. For $S = ½$, any site belongs only to a single bond in Λ_α. If k and l do not belong to the same bond, $|\alpha\rangle$ does not contain any correlation between $\vec{S}_k$ and $\vec{S}_l$ and then

$$\langle\alpha|\vec{S}_k \cdot \vec{S}_l|\alpha\rangle - \langle\alpha|\vec{S}_k|\alpha\rangle \cdot \langle\alpha|\vec{S}_l|\alpha\rangle = 0. \tag{5.81}$$

If k and l belong to the same bond, we have for $k \in A$ and $l \in B$ (without loss of generality):

$$
\begin{aligned}
\langle\alpha|\vec{S}_k\cdot\vec{S}_l|\alpha\rangle &= \frac{\langle\downarrow|_l\langle\uparrow|_k-\langle\uparrow|_l\langle\downarrow|_k}{\sqrt{2}}\left(\frac{S_k^+S_l^-+S_k^-S_l^+}{2}+S_k^zS_l^z\right)\\
&\quad\times\frac{|\uparrow\rangle_k|\downarrow\rangle_l-|\downarrow\rangle_k|\uparrow\rangle_l}{\sqrt{2}}\\
&= \frac{1}{2}\langle\downarrow|_l\langle\uparrow|_k\,S_k^zS_l^z\,|\uparrow\rangle_k|\downarrow\rangle_l-\frac{1}{4}\langle\downarrow|_l\langle\uparrow|_k\,S_k^+S_l^-\,|\downarrow\rangle_k|\uparrow\rangle_l\\
&\quad-\frac{1}{4}\langle\uparrow|_l\langle\downarrow|_k\,S_k^-S_l^+\,|\uparrow\rangle_k|\downarrow\rangle_l-\frac{1}{2}\langle\uparrow|_l\langle\downarrow|_k\,S_k^zS_l^z\,|\downarrow\rangle_k|\uparrow\rangle_l\\
&= -\frac{1}{8}-\frac{1}{4}-\frac{1}{4}-\frac{1}{8}=-\frac{3}{4}.
\end{aligned}
\tag{5.82}
$$

So, we have the result

$$
\langle\alpha|\vec{S}_k\cdot\vec{S}_l|\alpha\rangle-\langle\alpha|\vec{S}_k|\alpha\rangle\cdot\langle\alpha|\vec{S}_l|\alpha\rangle=\begin{cases}3/4 & \text{if } k=l\\ -3/4 & \text{if } (kl)\in\Lambda_\alpha \text{ or } (lk)\in\Lambda_\alpha.\\ 0 & \text{otherwise}\end{cases}
\tag{5.83}
$$

If all bonds in Λ_α are of short range. i.e., if there is a length ξ so that Λ_α does not contain bonds (ij) with $|\vec{r}_i-\vec{r}_j|\geqslant\xi$, the spin correlation in $|\alpha\rangle$ is of short range. Then any superposition of such states also has only a short-range spin correlation. States of this type are spin-liquid states. Note that, although any state $|\alpha\rangle$ breaks translational symmetry, the valence bond state $|\mathrm{VB}\rangle$ can restore this symmetry.

It is challenging to prove directly that an RVB state is realized for a given spin-1/2 model because it requires performing numerical simulations on enormous clusters.

Up to now, we have considered systems with fixed interactions. However, we can obtain impressive results in varying one or more parameters. For instance, let us consider the antiferromagnetic Heisenberg model on a square lattice shown in figure 5.4, with nearest-neighbor exchange interactions equal to J and J/λ with $\lambda\geqslant 1$. At $\lambda=1$ one has the isotropic Heisenberg model which, as we know, has a ground state with Néel order. For $\lambda\gg 1$, the ground state can be considered to be a product of nearest-neighbor singlet valence bonds on the J links, and this state is not smoothly connected to the Néel state. Quantum Monte Carlo simulations have shown a direct quantum phase transition between the two states at a critical value of λ (quantum phase transitions will be studied in section 6.2).

5.9 VBS ground states for spins larger than 1/2

One way to construct a VBS ground state with spin $S=1$ is to associate two spins ½ at each lattice site and create the spin one state by symmetrizing them [15, 16]. To avoid the formation of spin $S=2$, we antisymmetrize states between different neighbor lattice sites. The graph representing the VBS ground state in one dimension is presented in figure 5.5. In the same way, we can get VBS states for higher values of

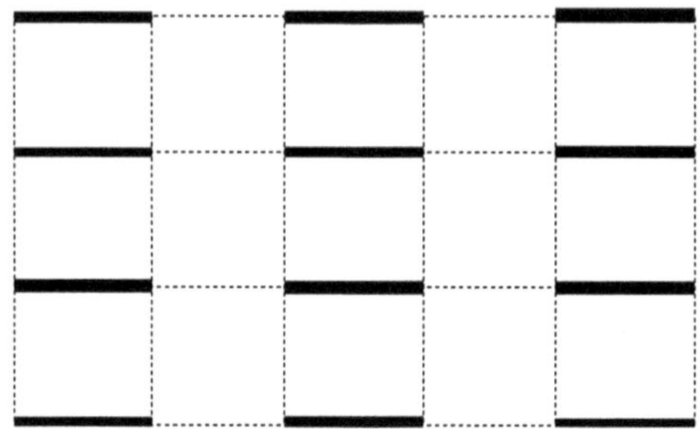

Figure 5.4. The full lines represent the exchange interaction J. The dashed lines represent the exchange J/λ.

$S = 1$

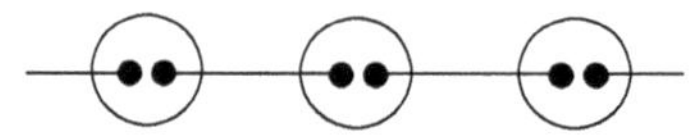

Figure 5.5. Small solid circles are spins ½. Lines represent antisymmetrization and circles symmetrization.

$S = 2$

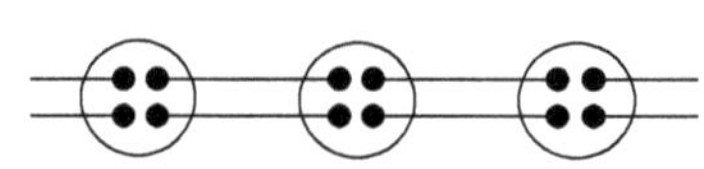

Figure 5.6. The graph for spin $S = 2$.

spins. In figure 5.6 I show the graph for $S = 2$. In terms of Schwinger bosons, the VBS ground state is given by

$$|\text{VBS}\rangle = \prod_{i=0}^{L} \left(a_i^\dagger b_{i+1}^\dagger - b_i^\dagger a_{i+1}^\dagger\right)^S |0\rangle. \tag{5.84}$$

5.10 Fermion operators

In this section, I introduce a SU(2) fermionic representation. The fermionic representation does not display magnetically ordered phases and tends to favor dimerized states. As in the case of Schwinger bosons, two spin ½ charge-neutral spinon operators $f_{i\alpha}$ ($\alpha = 1, 2$) are introduced. These operators satisfy anticommutation relations, and the spin operator $\vec{S}_i$ is represented by [17]

$$\vec{S}_i = \frac{1}{2} f_{i\alpha}^\dagger \vec{\sigma}_{\alpha\beta} f_{i\beta}, \tag{5.85}$$

or

$$S_i^+ = f_{i\alpha}^\dagger f_{i\beta}, \qquad S_i^- = f_{i\beta}^\dagger f_{i\alpha}, \qquad S_i^z = \frac{1}{2}\left(f_{i\alpha}^\dagger f_{i\alpha} - f_{i\beta}^\dagger f_{i\beta}\right). \tag{5.86}$$

The Hamiltonian

$$H = \sum_{\langle ij \rangle} J_{ij} \vec{S}_i \cdot \vec{S}_j, \tag{5.87}$$

is then written as

$$H = -\frac{1}{2}\sum_{\langle ij\rangle} J_{ij} f_{i\alpha}^{\dagger} f_{j\alpha} f_{j\beta}^{\dagger} f_{i\beta} + \sum_{\langle ij\rangle} J_{ij}\left(\frac{1}{2}n_i - \frac{1}{4}n_i n_j\right). \tag{5.88}$$

The second term in (5.88) is a constant and can be dropped. As before, we impose the constraint

$$f_{i1}^{\dagger} f_{i1} + f_{12}^{\dagger} f_{i2} = 1(= 2S). \tag{5.89}$$

This representation is limited to $S = 1/2$ due to the Pauli principle. We obtain a mean-field ground state at zero-order using the following approximations. First, we impose the constraint on the average using a site-dependent and time-independent Lagrange multiplier $\lambda_i(f_{i\alpha}^{\dagger} f_{i\alpha} - 1)$. Second, we replace the operator $f_{i\alpha}^{\dagger} f_{j\alpha}$ by its ground-state expectation value $\langle f_{i\alpha}^{\dagger} f_{j\alpha}\rangle = \chi_{ij}$. The MF Hamiltonian is then given by

$$H_{MF} = -\frac{1}{2}\sum_{\langle ij\rangle} J_{ij}\left[\left(f_{i\alpha}^{\dagger} f_{j\alpha}\chi_{ji} + h.c.\right) - |\chi_{ij}|^2\right] + \sum_{i}\lambda_i\left(f_{i\alpha}^{\dagger} f_{i\alpha} - 1\right). \tag{5.90}$$

The excitations in the zeroth-order mean-field theory are free spinons described by $f_{i\alpha}$. The term χ_{ij} has two kinds of fluctuations, amplitude fluctuations and phase fluctuations. The amplitude fluctuations can be neglected since they have a finite energy gap. The phase fluctuation a_{ij} around the mean-field ansatz $\bar{\chi}_{ij}$ can be taken into account writing

$$\chi_{ij} = \bar{\chi}_{ij} e^{-ia_{ij}}. \tag{5.91}$$

We can then write the mean-field Hamiltonian as

$$H = -\sum_{\langle ij\rangle} J_{ij}\left(f_{i\alpha}^{\dagger} f_{j\alpha}\bar{\chi}_{ji} e^{-ia_{ji}} + h.c.\right) - \sum_{i,\alpha}\lambda_i\left(f_{i\alpha}^{\dagger} f_{i\alpha} - 1\right), \tag{5.92}$$

where we have included fluctuations in λ. Equation (5.92) was called the first-order mean-field Hamiltonian by Wen [17]. The Hamiltonian is invariant under the gauge transformation:

$$a_{ij} \to a_{ij} + \theta_i - \theta_j, \qquad f_i \to f_i e^{i\theta_i}, \tag{5.93}$$

and the spinons are coupled to a U(1) gauge field (λ_i, a_{ij}). The original Heisenberg model was an interacting spin model with boson excitations. In the above procedure, the bosonic spin operator was split into a product of two fermionic operators. Then a gauge field was introduced to glue the spinons back into a bosonic spin. It is expected then that the first-order mean-field theory could work if the gauge field is in a confining phase, where the spinons interact with each other through a linear potential and never appear as free quasiparticles at low energies.

As an example, I will discuss the π-flux state (defined in section 5.5) for the nearest-neighbor Heisenberg model. This state is given by the ansatz [17]:

$$\chi_1 = \chi_{j,j+\hat{e}_x}, \quad \chi_2 = \chi_{j+\hat{e}_x,j+\hat{e}_y}, \quad \chi_3 = \chi_{j+\hat{e}_x+\hat{e}_y,j+\hat{e}_y}, \quad \chi_4 = \chi_{j+\hat{e}_y,j}, \qquad \lambda_i = 0. \tag{5.94}$$

We have two nonequivalent sites on sublattices A and B. Let us write $\bar{f}_j$ when $j \in B$. The Hamiltonian can be written as

$$\begin{aligned} H = &- J\sum_{j\in A}\sum_{\alpha}\left(\chi_1 \bar{f}^{\dagger}_{j+\hat{e}_x,\,\alpha} f_{j\alpha} + \chi_4^* \bar{f}^{\dagger}_{j+\hat{e}_y} f_{j\alpha} + h.c.\right) \\ &- J\sum_{j\in B}\sum_{\alpha}\left(\chi_3^* f^{\dagger}_{j+\hat{e}_x,\,\alpha} \bar{f}_{j\alpha} + \chi_2 f^{\dagger}_{j+\hat{e}_y,\,\alpha} \bar{f}_{j\alpha} + h.c.\right) \\ &+ JN\left(|\chi_1|^2 + |\chi_2|^2 + |\chi_3|^2 + |\chi_4|^2\right). \end{aligned} \tag{5.95}$$

Taking

$$\chi_1 = |\chi|, \quad \chi_2 = i|\chi|, \quad \chi_3 = |\chi|, \quad \chi_4 = i|\chi|, \tag{5.96}$$

into H and Fourier transforming, we get

$$H = J\sum_k\sum_\alpha |\chi|\left\{f^{\dagger}_{k\alpha}\bar{f}_{k\alpha}(\cos k_x + i\cos k_y) + \bar{f}^{\dagger}_{k\alpha} f_{k\alpha}(\cos k_x - i\cos k_y)\right\}. \tag{5.97}$$

We can write

$$H = J\sum_{k,\alpha}\left(f^{\dagger}_{k\alpha}\ \bar{f}^{\dagger}_{k\alpha}\right)\begin{pmatrix} 0 & |\chi|(\cos k_x + i\cos k_y) \\ |\chi|(\cos k_y - i\cos k_y) & 0 \end{pmatrix}\begin{pmatrix} f_{k\alpha} \\ \bar{f}_{k\alpha} \end{pmatrix}, \tag{5.98}$$

and we get the spinon spectrum

$$E_{\pm}(k) = \pm J|\chi|\sqrt{\cos^2 k_x + \cos^2 k_y}. \tag{5.99}$$

The excitations are gapless Dirac fermions with spin ½ and electrically neutral. Because there are two inequivalent sublattices, the dispersion relation has two branches. We can see that the gap vanishes at $\vec{k} = (\pm\pi, \pm\pi)$.

The flux in each plaquette is $\theta_{12} + \theta_{23} + \theta_{34} + \theta_{41} = \pi \pmod{2\pi}$. Since $(i)^2 = -1 = e^{i\pi}$ we have a π-flux (figure 5.7). The function χ is obtained by minimizing the ground-state mean-field energy. To calculate the ground-state energy, we must sum the energies of the lower branch (since the band is half-filled) and then add the term quadratic in χ. That gives

$$E_g = -\sum_k J|\chi|\sqrt{\cos^2 k_x + \cos^2 k_y} + J\,|\chi|^2 N. \tag{5.100}$$

We have then

$$|\chi| = \frac{1}{4}\int\frac{d^2\vec{k}}{(2\pi)^2}\sqrt{\cos^2 k_x + \cos^2 k_y}. \tag{5.101}$$

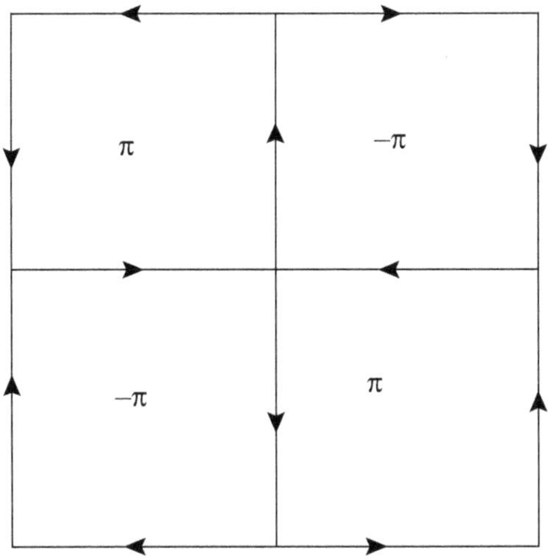

Figure 5.7. The mean-field ansatz of the π-flux state.

At the mean-field level, the π-flux state contains gapless spin excitations, which correspond to the particle-hole excitations. In addition to these excitations, the mean-field theory also has U(1) gauge fluctuations (a_{ij}, λ_i). The interaction between the spin excitations and gauge fluctuations is discussed in detail by Wen [17] in his excellent book, where also several ansatzes for spin liquids are presented.

5.11 Holons

Let us consider a lattice of dimers (each dimer has spin 0, so the ground state has spin 0). As was mentioned before, the elementary magnetic excitations involve replacing one of the singlets by a triplet state. Fractionalization of the triplet means that the two spin-1/2 excitations (spinons with charge zero) can propagate independently. If we remove a single electron, its removal results in breaking one of the dimers, leaving an empty state (called a *holon*) and a free spin (spinon) on the same bond. If the dimers can change their positions by quantum fluctuations, the holon and the spinon can move on the lattice. The electron will have fractionalized into the spinon carrying spin ½ and the holon with spin-zero and electric charge $+e$. This effect is called spin-charge separation.

We can define a set of Bose and Fermi operators b_i and $f_{\alpha i}$, at each site i, satisfying the constraint at each site [13]:

$$b_i^\dagger b_i + f_{\alpha i}^\dagger f_{\alpha i} = 1 \quad \text{(no sum in } i\text{)}. \tag{5.102}$$

Let $|\bar{0}\rangle$ be the reference state for these operators and define the states $|h\rangle, |\uparrow\rangle$ and $|\downarrow\rangle$ representing a holon with charge $+e$ and spin-zero and spinons $|\uparrow\rangle$ with spin up and $|\downarrow\rangle$ with spin down and zero charges:

$$|h\rangle \equiv |e, 0\rangle = b^\dagger\,|\bar{0}\rangle, \quad |\uparrow\rangle \equiv |0, \uparrow\rangle = f_\uparrow^\dagger\,|\bar{0}\rangle \quad |\downarrow\rangle \equiv |0, \downarrow\rangle = f_\downarrow^\dagger\,|\bar{0}\rangle. \tag{5.103}$$

The operator $c_{\sigma i}^{\dagger}$, which creates a fermion of charge e and spin σ at each site i, is given by

$$c_{\sigma i}^{\dagger} = b_i f_{\sigma i}^{\dagger} \text{ (no sum in } i). \tag{5.104}$$

5.12 The dimer order parameter

In this section, I will introduce the dimer order parameter for a one-dimensional spin 1/2 system described by the Hamiltonian [18]

$$H = \sum_{l=1}^{N}(J_1\vec{S}_l \cdot \vec{S}_{l+1} + J_2\vec{S}_l \cdot \vec{S}_{l+2}), \tag{5.105}$$

with $J_2/J_1 = 1/2$ and periodic boundary conditions. The system has a pure dimer order in the ground state

$$\left|\phi_1^{D}\right\rangle = [1, 2][3, 4]\dots[N-1, N], \qquad \left|\phi_2^{D}\right\rangle = [2, 3][4, 5]\dots[N, 1]. \tag{5.106}$$

The dimer order parameter is given by the operator

$$D = \frac{1}{N}\sum_{l=1}^{N}(-1)^l D_l, \qquad D_l = S_l^{+}S_{l+1}^{-} + S_l^{-}S_{l+1}^{+}. \tag{5.107}$$

Dimer order manifests itself in the four-spin correlation function $\langle D_l D_{l+n}\rangle$. We have

$$\langle D_l D_{l+n}\rangle - \langle D_l\rangle\langle D_{l+n}\rangle = \frac{1}{4}(-1)^n. \qquad n \neq 0. \tag{5.108}$$

We can generalize the dimer order parameter for two dimensions. In general, we cannot calculate this parameter analytically, but it is useful in numerical calculations [19].

5.13 The Shastry–Sutherland lattice

The Shastry–Sutherland lattice (figure 5.8) for spin ½ was studied by Shastry and Sutherland [20] as an example of a frustrated quantum spin system with an exact ground state. It is described as a square lattice with J' antiferromagnetic coupling between nearest neighbors and J antiferromagnetic couplings between next nearest-neighbors in every second square along the diagonals. In the classical limit, when $S \to \infty$ the ground state is Néel ordered if $J/J' \leqslant 1$ and is helically ordered otherwise, where the twist between one spin and its nearest neighbor is given by $q = \cos^{-1}(-J'/J)$ [21]. Shastry and Sutherland [20] have shown that the dimer state (built from singlets) is an exact eigenstate of the Hamiltonian for any J, J' and that it is the exact ground state for $J/J' > 2$. Numerical calculations have shown that for $J'/J \geqslant 1.55 \pm 0.05$ the model has a gap in the spectrum of excitations and is in a dimerized valence bond crystal. For $J'/J \leqslant 1.55 \pm 0.05$ the system exhibits Néel

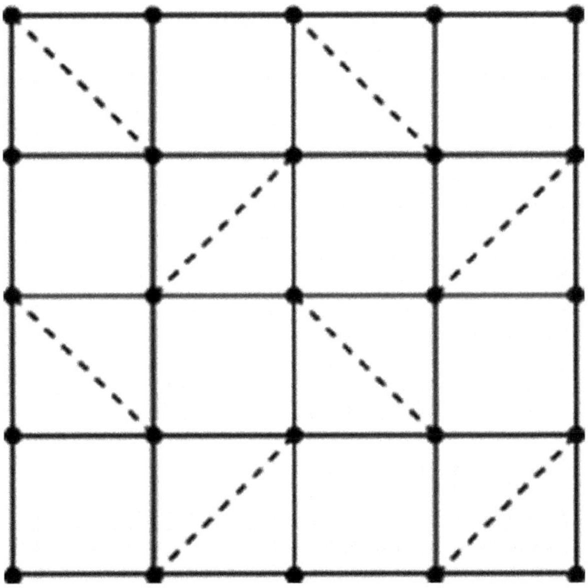

Figure 5.8. The exchange J' acts along the horizontal and vertical lines, while the exchange J acts along the diagonal dashed lines.

order. The possibility of an intermediate phase, possibly with short-range helical correlations, has been discussed in the literature [22].

5.14 Exercises

5.1. Show that the state (5.10) is an eigenstate of the total spin $\vec{S}^2$ and S^z operators with the correct eigenvalues $S(S + 1)$ and m, respectively.

5.2. Find the self-consistent equations for the ferrimagnetic chain, introduced in problem 2.1, using the mean-field Schwinger boson formalism (see [23]).

5.3. Find the self-consistent equations for the antiferromagnetic Heisenberg model on a triangular lattice in the presence of a magnetic field using the mean-field Schwinger boson formalism. Note: when we use the Schwinger boson technique, we do not need to rotate the axis in the presence of a magnetic field, as was done in the case of the spin-wave theory (see [24]).

5.4. The two-dimensional square lattice anisotropic Heisenberg antiferromagnet is described by the following Hamiltonian

$$h = J_x \sum_{\langle i,j \rangle} \vec{S}_i \cdot \vec{S}_j + J_y \sum_{\langle l,m \rangle} \vec{S}_l \cdot \vec{S}_m,$$

where the sums $\langle i, j \rangle$ and $\langle l, m \rangle$ are defined for the nearest neighbors along the x- and y-directions, respectively. Using the Schwinger boson mean-field theory find the self-consistent equations for the parameters (see [25]).

5.5. Consider the spin ½ Majundar–Ghosh Hamiltonian

$$H = \frac{4|J|}{3}\sum_{n=1}^{N}\left(\vec{S}_n \cdot \vec{S}_{n+1} + \frac{1}{2}\vec{S}_n \cdot \vec{S}_{n+2}\right) + \frac{N|J|}{2},$$

where the total number of sites N is even, and $\vec{S}_{N+1} = \vec{S}_1$. Show that the dimer (or valence bond) states:

$$|\Omega_\pm\rangle = \prod_{n=1}^{N/2} \frac{1}{\sqrt{2}}\left(|\uparrow\rangle_{2n} \otimes |\downarrow\rangle_{2n\pm1} - |\downarrow\rangle_{2n} \otimes |\uparrow\rangle_{2n\pm1}\right)$$

are exact ground states, i.e., $|\Omega_+\rangle$ describes the state where neighboring spins on sites $2n$ and $2n+1$ are in a total singlet ($S = 0$) state (note: write the Hamiltonian in terms of the total spin of a triad: $\vec{J}_n = \vec{S}_{n-1} + \vec{S}_n + \vec{S}_{n+1}$) (see [1])).

References

[1] Auerbach A 1994 *Interacting Electrons and Quantum Magnetism* (Berlin: Springer)

[2] Mezio A, Sposeti C N, Manuel L O and Trumper A E 2011 A test of the bosonic spinon for the triangular antiferromagnet spectrum *Eur. Phys. Lett.* **94** 47001

[3] Ceccato H A, Gaza C J and Trumper A E 1993 Nonclassical disordered phase in the strong quantum limit of frustrated antiferromagnets *Phys. Rev.* B 12329–32

[4] Mila F, Poilblanc D and Bruder C 1991 Spin dynamics in a frustrated magnet with short range-order *Phys. Rev.* B **43** 7891–8

[5] Flint R and Coleman P 2009 Symplectic N and time reversal in frustrated magnetism *Phys. Rev.* B **79** 014424

[6] Read N and Sachdev S 1991 Large-N expansion for frustrated quantum antiferromagnets *Phys. Rev. Lett.* **66** 1773–6

[7] Yoshioka D 1989 Boson mean-field theory of the square lattice Heisenberg model *J. Phys. Soc. Jpn.* **58** 3733–45

[8] Schmidt B and Thalmeier P 2017 Frustrated two-dimensional quantum magnets *Phys. Rep.* **703** 1–59

[9] Misguich G and Lhuillier C 2005 Two dimensional quantum antiferromagnets *Frustrated Spin Systems* ed H T Diep (Singapore: World Scientific)

[10] Mila F 2015 Frustrated spin systems *Many-Body Physics: From Kondo to Hubbard* ed E Pavarini, E Koch and P Coleman (Julich: Forschungszentrum Julich) ch 7

[11] Yang X and Wang F 2016 Schwinger boson spin liquid states on a square lattice *Phys. Rev.* B **94** 035160

[12] Lhuillier C and Misguich G 2011 Introduction to quantum spin liquids *Introduction to Frustrated Magnetism* ed C Lacroix, P Mendels and F Mila (Heidelberg: Springer) pp 23–41

[13] Fradkin E 1991 *Field Theories of Condensed Matter Systems* (Redwood City, CA: Addison-Wesley Publishing Company)

[14] Timm C 2011 *Theory of Magnetism* (Dresden: Technische Universitat Dresden)

[15] Santos R A and Korepin V 2012 Entanglement of disjoint blocks in the one-dimensional spin 1 *J. Phys. A: Math.Theor.* **45** 125307

[16] Affleck I, Kenedy T, Lieb E H and Tasaki H 1987 Rigorous results on valence-bond ground states in antiferromagnets *Phys. Rev. Lett.* **59** 799–802

[17] Wen X G 2004 *Quantum Field Theory of Many-Body Systems* (Oxford: Oxford University Press)

[18] Yu Y, Muller G and Viswanath V S 1996 Dimer and Néel order-parameter fluctuations in the spin-fluid phase of the s = 1/2 spin chain with first- and second-neighbor couplings *Phys. Rev.* B **54** 9242–9

[19] Sandvik A W 2007 Evidence for deconfined quantum criticality in a two-dimensional Heisenberg model with four-spin interactions *Phys. Rev. Lett.* **98** 227202

[20] Shastry B S and Sutherland B 1981 Exact ground state of a quantum mechanical antiferromagnet *Physica* B **108** 1069–70

[21] Weihong Z, Oitmaa J and Hamer C J 2001 Phase diagram of the Shastry–Sutherland antiferromagnet *Phys. Rev.* B **65** 014408

[22] Lauchli A, Wessel S and Sigrist M 2002 Phase diagram of the quadrumerized Shastry–Sutherland model *Phys. Rev.* B **66** 014401

[23] Wu C, Chen B, Dai X, Yu Y and Su Z B 1999 Schwinger boson mean-field theory of the Heisenberg ferrimagnetic spin chain *Phys. Rev.* B **60** 1057–63

[24] Chen S Q and Zhang F C 2002 Antiferromagnetic Heisenberg model on an anisotropic triangular lattice in the presence of a magnetic field *Phys. Rev.* B **66** 172407

[25] Ji Y and Dong J 2003 Schwinger-boson mean field solutions of spin-1 (or ½) 2D anisotropic Heisenberg antiferromagnet arXiv:cond-mat/0301570

Antonio Sergio Teixeira Pires

Chapter 6

Bond operators and Schwinger SU(3) bosons

In the last chapter, I discussed the SU(2) Schwinger boson formalism, a quite adequate tool to treat magnetically disordered phases. In this chapter, I introduce two new kinds of operators: The bond operators to handle dimerized phases and the SU(3) Schwinger boson operators, an extension of the SU(2) formalism, to treat nematic phases in spin-one systems.

6.1 Bond operators

Sachdev and Bhatt [1] introduced a bond operator representation of quantum spin ½ that was specifically designed to treat the properties of dimerized phases. The method applies to both spontaneously and intrinsically dimerized phases. To introduce the technique, I use two $S = ½$ spins $\vec{S}_1$ and $\vec{S}_2$. The four states in the Hilbert space are combined to form the singlet state $|s\rangle$ and the triplet states $|t_x\rangle$, $|t_y\rangle$, and $|t_z\rangle$. Four bosonic operators that create these states are introduced as

$$\begin{aligned}
|s\rangle &\equiv s^\dagger\,|0\rangle = \frac{1}{\sqrt{2}}(|\uparrow\downarrow\rangle - |\downarrow\uparrow\rangle),\\
|t_x\rangle &\equiv t_x^\dagger\,|0\rangle = -\frac{1}{\sqrt{2}}(|\uparrow\uparrow\rangle - |\downarrow\downarrow\rangle),\\
|t_y\rangle &\equiv t_y^\dagger\,|0\rangle = \frac{i}{\sqrt{2}}(|\uparrow\uparrow\rangle + |\downarrow\downarrow\rangle),\\
|t_z\rangle &\equiv t_z^\dagger\,|0\rangle = \frac{1}{\sqrt{2}}(|\uparrow\downarrow\rangle + |\downarrow\uparrow\rangle),
\end{aligned} \tag{6.1}$$

where $|0\rangle$ is the vacuum. For physical states, there is either a singlet or triplet, so the following constraints should be imposed

doi:10.1088/978-0-7503-3879-0ch6

$$S^x|\uparrow\rangle = \frac{1}{2}|\downarrow\rangle, \;\; S^x|\downarrow\rangle = \frac{1}{2}|\uparrow\rangle,$$
$$S^y|\uparrow\rangle = \frac{i}{2}|\downarrow\rangle, \;\; S^y|\downarrow\rangle = \frac{-i}{2}|\uparrow\rangle, \tag{6.2}$$
$$S^z|\uparrow\rangle = \frac{1}{2}|\uparrow\rangle, \;\; S^z|\downarrow\rangle = -\frac{1}{2}|\downarrow\rangle.$$

Applying the component S_1^x to the state $|s\rangle$ we get

$$S_1^x|s\rangle = S_1^x\left[\frac{1}{\sqrt{2}}(|\uparrow\downarrow\rangle - |\downarrow\uparrow\rangle)\right] = \frac{1}{2}\left[\frac{1}{\sqrt{2}}(|\downarrow\downarrow\rangle - |\uparrow\uparrow\rangle)\right] = -\frac{1}{2}\left[\frac{1}{\sqrt{2}}(|\uparrow\uparrow\rangle - |\downarrow\downarrow\rangle)\right]$$
$$= \frac{1}{2}|t_x\rangle.$$

So

$$S_1^x|s\rangle = S_1^x s^\dagger|0\rangle = \frac{1}{2}t_x.$$

However, $\frac{1}{2}t_x^\dagger s s^\dagger|0\rangle = \frac{1}{2}|t_x\rangle$, since $ss^\dagger|0\rangle = |0\rangle$, which leads to $S_1^x \propto \frac{1}{2}t_x^\dagger s$.

Consider now

$$S_1^x|t_x\rangle = S_1^x\left[-\frac{1}{\sqrt{2}}(|\uparrow\uparrow\rangle - |\downarrow\downarrow\rangle)\right] = \frac{1}{2}|s\rangle.$$

Hence, we can write

$$S_1^x|t_x\rangle = S_1^x t_x^\dagger|0\rangle = \frac{1}{2}|s\rangle.$$

However, $\frac{1}{2}s^\dagger t_x t_x^\dagger|0\rangle = \frac{1}{2}|s\rangle$. So $S_1^x \propto \frac{1}{2}s^\dagger t_x$.

Applying now S_1^x to the state $|t_y\rangle$ we obtain

$$S_1^x|t_y\rangle = S_1^x\left[\frac{i}{\sqrt{2}}(|\uparrow\uparrow\rangle + |\downarrow\downarrow\rangle)\right] = \frac{i}{2}|t_z\rangle.$$

Writing $S_1^x|t_y\rangle = S_1^x t_y^\dagger|0\rangle = \frac{i}{2}|t_z\rangle$, and $\frac{i}{2}t_z^\dagger t_y t_y^\dagger|0\rangle = \frac{i}{2}|t_z\rangle$, we get $S_1^x \propto \frac{i}{2}t_z^\dagger t_y$.

Finally, we have

$$S_1^x|t_z\rangle = S_1^x\left[\frac{1}{\sqrt{2}}(|\uparrow\downarrow\rangle + |\downarrow\uparrow\rangle)\right] = -\frac{i}{2}|t_y\rangle.$$

As before, we use the expression $\frac{i}{2}t_y^\dagger t_z t_z^\dagger|0\rangle = \frac{i}{2}|t_y\rangle$, to conclude that $S_1^x \propto -\frac{i}{2}t_y^\dagger t_z$.

The above calculations allow us to write

$$S_1^x = \frac{1}{2}\Big[s^\dagger t_x + t_x^\dagger s - i\big(t_y^\dagger t_z - t_z^\dagger t_y\big)\Big]. \tag{6.3}$$

For the S_1^y component, we find

$$S_1^y = \frac{1}{2}\Big[s^\dagger t_y + t_y^\dagger s + i(t_x^\dagger t_z - t_z^\dagger t_x)\Big]. \tag{6.4}$$

We can write the final result in the form

$$\begin{aligned} S_1^\alpha &= \frac{1}{2}\big(s^\dagger t_\alpha + t_\alpha^\dagger s - i\varepsilon_{\alpha\beta\gamma} t_\beta^\dagger t_\gamma\big), \\ S_2^\alpha &= \frac{1}{2}\big(-s^\dagger t_\alpha - t_\alpha^\dagger s - i\varepsilon_{\alpha\beta\gamma} t_\beta^\dagger t_\gamma\big), \end{aligned} \tag{6.5}$$

where ε is the antisymmetric tensor and α, β, γ take the values x, y, z, and repeated indices are summed over. The restriction that the physical states are either singlets or triplets leads to the constraint

$$s^\dagger s + t_\alpha^\dagger t_\alpha = 1. \tag{6.6}$$

If we consider the operators as bosonic, they satisfy the commutation relations

$$[s, s^\dagger] = 1, \;\; [t_\alpha, t_\beta^\dagger] = \delta_{\alpha\beta}, \;\; [s, t_\alpha^\dagger] = 0. \tag{6.7}$$

The bond operator formalism yields two different classes of ground states:

(i) *Dimerized phases*, which are magnetically disordered states. The dimerization can be either spontaneous or intrinsic. For those states, we have

$$\langle s\rangle \neq 0, \;\; \langle t_\alpha\rangle = 0, \;\; \langle t_\alpha t_\beta\rangle \; = c\delta_{\alpha\beta}, \tag{6.8}$$

where c is a non-zero constant. The t_α bosons have condensed in pairs, but there is no single t_α boson condensation.

(ii) *Magnetically ordered phases.* Condensation of single t_α bosons leads to long-range magnetic order. We have $\langle s\rangle \neq 0$, $\langle t_\alpha\rangle \neq 0$.

The wavevector and polarization of the mode at which the t_α bosons condense determine the nature of the magnetic order.

As an example, we treat the bilayer Heisenberg model studied by Matsushita *et al* [2] described by the Hamiltonian

$$H = J\left[\sum_{\langle ij\rangle}\vec{S}_{1,i}\cdot\vec{S}_{1,j} + \sum_{\langle ij\rangle}\vec{S}_{2,i}\cdot\vec{S}_{2,j}\right] + J_2\sum_i \vec{S}_{1,i}\cdot\vec{S}_{2,i}. \tag{6.9}$$

Using the bond operator expressions (6.5), we write H as

$$
\begin{aligned}
H = J_2\sum_i\left(-\frac{3}{4}s_i^\dagger s_i + \frac{1}{4}t_{\alpha i}^\dagger t_{\alpha i}\right) - \sum_i \mu_i(s_i^\dagger s_i + t_{\alpha i}^\dagger t_{\alpha i} - 1) \\
+ \frac{J}{2}\sum_{\langle ij\rangle}\left(s_i^\dagger s_j^\dagger t_{\alpha i}t_{\alpha j} + s_i^\dagger s_j t_{\alpha i}t_{\alpha j}^\dagger + \text{h. c.}\right) \\
- \frac{J}{2}\sum_{\langle ij\rangle}(1 - \delta_{\alpha\beta})\left(t_{\alpha i}^\dagger t_{\beta j}t_{\alpha i}^\dagger t_{\beta j} - t_{\alpha i}^\dagger t_{\beta j}t_{\beta i}^\dagger t_{\alpha j}\right),
\end{aligned}
\tag{6.10}
$$

where the Lagrange multiplier μ_i enforces the constraint (6.6). In a mean-field approximation, we take $\mu = \mu_i$ and condense the singlet, i.e. $\langle s_i^\dagger\rangle = \langle s_i\rangle = s$. Neglecting high order terms, we diagonalize the Hamiltonian using the Bogoliubov transformation

$$
\xi_{k\alpha} = \cosh\theta_k t_{k\alpha} + \sinh\theta_k t_{-k\alpha}^\dagger. \tag{6.11}
$$

We find

$$
\cosh^2\theta_k = \frac{1}{2}\left(\frac{\Lambda_k}{\omega_k} + 1\right), \tag{6.12}
$$

and

$$
H_{MF} = N\left(-\frac{3}{4}s^2 - \mu s^2 + \frac{5}{2}\mu - \frac{3}{8}\right) + \sum_k \omega_k\left(\xi_{k\alpha}^\dagger\xi_{k\alpha} + \frac{3}{2}\right), \tag{6.13}
$$

where

$$
\omega_k = \sqrt{\Lambda_k^2 - 4\Delta_k^2},\ \Lambda_k = \frac{J_2}{4} - \mu + 2Js^2\gamma_k,\ \Delta_k = Js^2\gamma_k. \tag{6.14}
$$

The above mean-field description of the disordered phase will be valid only at low temperatures. We obtain self-consistent equations for the mean-field parameter s and μ by minimizing the total free energy. For details, see [2].

The bond operator method was generalized by Shik *et al* [3] to spins $S \geqslant 1$.

6.2 Quantum phase transitions

Quantum phase transitions (QPTs) occur, in a strict sense, at precisely zero temperature only. Such phase transitions are not controlled by changing the temperature, but rather some system parameters, such as pressure, magnetic field, anisotropies, or chemical composition. In contrast to phase transition at finite temperatures, quantum effects play a decisive role, as the fluctuations driving the transition follow quantum instead of classical statistical mechanics. The fluctuations that destroy the long-range order in the QPT are quantum fluctuations, which are a consequence of Heisenberg's uncertainty principle.

To be more specific, let us consider a Hamiltonian given by

$$
H(g) = H_1 + gH_2. \tag{6.15}
$$

A quantum phase transition occurs when we change the coupling parameter g, and the quantum ground state of the system changes in some fundamental way at a critical value of g called g_c. The parameter g transforms the system between quantum mechanical states that either look different topologically or have distinctly different magnetic properties. In any Hamiltonian that admits a QPT must be two non-commuting operators that describe two competing ordering tendencies. For instance, suppose that

$$H_1 = -\sum_{\langle i,j \rangle} \hat{n}_i \cdot \hat{n}_j \tag{6.16}$$

where $\hat{n} = (\cos \phi, \sin \phi)$. The ordered state minimizes the energy. Now we add a kinetic energy term given by

$$H_2 = \sum_i (L_i^z)^2, \tag{6.17}$$

where L_i^z is the z-component of the angular momentum at site i. This energy is minimized when the orientation is maximally uncertain (by the uncertainty principle). So, we have a disordered state. The competition between the two terms leads to the QPT.

The quantum critical point (QCP) can be viewed, in some cases, as the endpoint of a line of finite-temperature transitions. At the QCP, quantum fluctuations exist on all length scales and, therefore, can be observed at finite temperature. The typical time scale for a decay of the fluctuations is called the *correlation time* τ_c. As the critical point is approached, the correlation time diverges as $\tau_c \propto \xi^z$, where ξ is the correlation length, and z is the dynamical critical exponent. An important characteristic of QPT is that in the quantum critical regime, quasi-particle excitations are not well defined. For a detailed study of QPT, the reader should consult the excellent book by Sachdev [4].

In the following, we will study a simple model that shows a QPT. The model is described by the quantum spin-one two-dimensional Heisenberg antiferromagnet with a single-ion anisotropy:

$$H = J\sum_{ij} \vec{S}_i \cdot \vec{S}_j + D\sum_i (S_i^z)^2. \tag{6.18}$$

Note that individual spin-1 particles can display more types of order than their spin-1/2 counterparts. For example, every spin-1/2 single-particle state (assuming it is a pure state) has a fixed magnetization of exactly ½ (setting $\hbar = 1$). This is a consequence of the fact that any spin-1/2 single-particle state can be obtained by applying a specific rotation to the up-state in the z-basis (or any other reference state for that matter). By contrast, spin-1 particles can have a magnetization of anywhere between 0 and 1. Even in the absence of magnetization, the $|0\rangle$ state in the S^z–basis ($m = 0$) breaks spin-rotation symmetry, because $\langle (S^z)^2 \rangle = 0$, whereas the fluctuations in x- and y-directions are non-zero: $\langle (S^x)^2 \rangle = \langle (S^y)^2 \rangle \neq 0$.

The spectrum of the Hamiltonian (6.18) changes drastically as D varies from very small to very large values [5, 6]. For very large D the energy is minimized for $S_i^z = 0$. The large D phase, $D > D_C$, consists of a unique ground state with total magnetization $S^Z_{total} = 0$ separated by a gap from the first excited states, which lie in the sectors $S^z_{total} = \pm 1$. The primary excitation in this phase is a gapped $S = 1$ exciton with an infinite lifetime at zero temperature. For $0 < D < D_C$ the spins will order antiferromagnetically along some direction in the xy-plane at $T = 0$. Here I will be concerned only with the paramagnetic phase (i.e., $D > D_C$).

The paramagnetic phase is a quadrupole state, where only a spin-nematic order parameter $Q = \langle (S^z)^2 - 2/3 \rangle$ is non-zero. The nematic order is, in general, characterized by a non-zero spin quadrupole tensor $Q_{ij} \equiv \langle S_i^\alpha S_j^\beta + S_j^\beta S_i^\alpha \rangle$ which deviates from isotropy. In the case of spin ½, the spin values are $S^z = \pm 1/2$, and in a magnetically disordered ground state, the spins should pair in singlets. For $S = 1$, we can have $S^z = 0$ on all sites. The magnetization vanishes, but the system is ordered. That is, we have $\langle \vec{S}_i \rangle = 0$, but the rotational symmetry, as was said before, is broken since $\langle (S^x)^2 \rangle = \langle (S^y)^2 \rangle \neq \langle (S^z)^2 \rangle$. This phase is called nematic in an allusion to liquid crystals, where, in the nematic phase, the molecules have no positional order. Still, they self-align to have long-range directional order with their long axes roughly parallel. A spin-nematic state is a state that breaks spin-rotational symmetry but preserves time-reversal symmetry. Note that in a genuinely disordered phase, we have

$$\langle (S^z)^2 \rangle = \langle (S^x)^2 \rangle = \langle (S^y)^2 \rangle = S(S+1)/3. \tag{6.19}$$

Remember that in the isotropic Heisenberg model, the existence of a gap implied that the ground state was magnetically disordered. However, here the anisotropic term breaks the SU(2) symmetry, and we have nematic order with a gap.

6.3 Schwinger SU(3) bosons

It is impossible to describe the spin-nematic phase using the standard SU(2) Schwinger boson (SB) formalism as a boson condensate because a non-zero condensate of $\langle a \rangle$ or $\langle b \rangle$ necessarily produces a non-zero magnetic dipole moment. Papanicolaou [7] developed a theory averting this shortcoming of the SB formalism. His approach is a generalization SU(3) of the SU(2) Schwinger boson mean-field theory. A boson condensate can describe both magnetic orders and spin-nematic order in this construction. This formalism is sometimes called *the flavor wave theory*.

In this approach, we start by introducing the following time-reversal invariant basis of the SU(3) fundamental representation:

$$|x\rangle = i(|1\rangle - |-1\rangle)/\sqrt{2}, \;\; |y\rangle = (|1\rangle + |-1\rangle)/\sqrt{2}, \;\; |z\rangle = -i\,|0\rangle, \tag{6.20}$$

where $|n\rangle$ are eigenstates of S^z. The invariance under time-reversal T follows from the fact that T interchanges $|1\rangle$ and $|-1\rangle$, T $|0\rangle = -|0\rangle$, and T changes the sign of i. We introduce a set of three boson operators t_α ($\alpha = x, y, z$) defined by

$$t_x^\dagger\,|v\rangle = |x\rangle,\quad t_y^\dagger\,|v\rangle = |y\rangle,\quad t_z^\dagger\,|v\rangle = |z\rangle, \tag{6.21}$$

where $|v\rangle$ is the vacuum state, with the constraint

$$t_x^\dagger t_x + t_y^\dagger t_y + t_z^\dagger t_z = 1, \tag{6.22}$$

for single-site occupancy on each site. This constraint is called the hard-core condition, even if the particles are bosons. In terms of the t operators, we can write

$$S^\alpha = -i\varepsilon_{\alpha\beta\gamma} t_\beta^\dagger t_\gamma. \tag{6.23}$$

The elements of the basis (6.20) are zero-eigenvalues eigenstate of the corresponding spin operators: $S^x|x\rangle = S^y|y\rangle = S^z|z\rangle = 0$. We may choose the operators t's as either bosons or fermions. In principle, the bosons condense to the lowest energy state at low temperatures and form a quantum-ordered state. In contrast, the fermions tend to develop a Fermi sea and a quantum-disordered state. It can be verified that

$$[S^\alpha, S^\beta] = i\varepsilon_{\alpha\beta\gamma} S^\gamma. \tag{6.24}$$

The states $t_x^\dagger|v\rangle$ and $t_y^\dagger|v\rangle$ both consist of $S^z = \pm 1$ eigenstates and have the average $\langle S^z\rangle = 0$. This property will preserve the ground state's disorder.

We can also use the usual spin S^z basis, and to do so, we introduce two bosonic operators $u^\dagger$ and $d^\dagger$ given by [5]

$$u^\dagger = -\frac{1}{\sqrt{2}}\left(t_x^\dagger + it_y^\dagger\right),\ d^\dagger = \frac{1}{\sqrt{2}}\left(t_x^\dagger - it_y^\dagger\right), \tag{6.25}$$

and so

$$|1\rangle = u^\dagger\,|v\rangle,\quad |0\rangle = t_z^\dagger\,|v\rangle,\quad |-1\rangle = d^\dagger\,|v\rangle, \tag{6.26}$$

with the constraint $u^\dagger u + d^\dagger d + t_z^\dagger t_z = 1$. The spin operators are now written as

$$S^+ = \sqrt{2}(t_z^\dagger d + u^\dagger t_z),\ S^- = \sqrt{2}(d^\dagger t_z + t_z^\dagger u),\ S^z = u^\dagger u - d^\dagger d. \tag{6.27}$$

Depending on the nature of the order that we consider (either a magnetic or quadrupolar order), we choose either of the above representations.

Since $u_i^\dagger u_i$, $d_i^\dagger d_i$, $t_{zi}^\dagger t_{zi}$ are 0 or 1, the single-site occupancy implies $d_i^\dagger d_i u_i^\dagger u_i = 0$. That leads to $(S_i^z)^2 = u_i^\dagger u_i + d_i^\dagger d_i$.

Writing the operators t's as a vector $\vec{t} = (t_x, t_y, t_z)^\mathrm{T}$ we have

$$\vec{S}_n = -i\vec{t}_n^{\,\dagger} \times \vec{t}_n. \tag{6.28}$$

This representation has a U(1) gauge freedom $\vec{t} \to e^{i\theta}\vec{t}$. We remark that $\vec{t}$ is not a spin-1/2 representation of spin SU(2) symmetry.

We have thus extended the SU(2) Schwinger boson formalism to SU(3). But note that this formalism is different from the large-N Schwinger boson mean-field theory introduced by Arovas and Auerbach [8] and discussed in chapter 5.

Now, besides the three components of spin $S_r^\alpha(\alpha = x, y, z)$, one has the following operators (quadrupolar order parameters) that describe quadrupolar order:

$$\begin{aligned}
Q_n^{(0)} &= (S_n^z)^2 - \frac{2}{3} = \frac{1}{3}\left(t_{nx}^\dagger t_{nx} + t_{ny}^\dagger t_{ny} - 2t_{nz}^\dagger t_{nz}\right),\\
Q_n^{(2)} &= (S_n^x)^2 - (S_n^y)^2 = -\left(t_{nx}^\dagger t_{nx} - t_{ny}^\dagger t_{ny}\right),\\
Q_n^{xy} &= S_n^x S_n^y + S_n^y S_n^x = -\left(t_{nx}^\dagger t_{ny} + t_{ny}^\dagger t_{nx}\right),\\
Q_n^{yz} &= S_n^y S_n^z + S_n^z S_n^y = -\left(t_{ny}^\dagger t_{nz} + t_{nz}^\dagger t_{ny}\right),\\
Q_n^{zx} &= S_n^z S_n^x + S_n^x S_n^z = -(t_{nz}^\dagger t_{nx} + t_{nx}^\dagger t_{nz}).
\end{aligned} \tag{6.29}$$

The three components of S_n^α (for any value of the spin S) obey the standard commutation rule and are, therefore, generators of the SU(2) representation. However, if we want to consider the ordering described by the operators (6.21), we should enlarge the symmetry group. We now have eight parameters (S_n^α, Q), which are generators of the SU(3) representation. The same happens for higher values of spins. For instance, for $S = 3/2$, one has octupolar order, and thus to treat this order, we should use a SU(4) representation.

The nematic order parameters $Q^{\alpha\beta}$ describe the anisotropy of spin fluctuations, not static moment, and can be non-zero only if $S \geqslant 1$. The paramagnetic phase has a net nematic component induced by the single-ion anisotropy, but no net magnetization component. The large D phase is magnetically disordered, but it has a nematic order since the nematic order parameter is non-null $\langle (Q_n^0) \rangle \neq 0$. Here, as I said before, I will study only the nematic phase of the Hamiltonian (6.18). A natural generalization of the spin-wave theory is obtained by extending the Holstein–Primakoff representation from SU(2) to SU(3). This generalization can be made by condensing one of the three SB: the corresponding creation and annihilation operators are replaced by a number according to the constraint (6.22) [9]. The SB that is condensed is the one that creates the local state that minimizes the mean-field energy. In the case of nematic order, the t_z boson is the one that is condensed. Using the constraint (6.22), we get

$$t_{zi}^\dagger = t_{zi} = \sqrt{1 - u_i^\dagger u_i - d_i^\dagger d_i}\,. \tag{6.30}$$

Another approach is to replace equation (6.28) by $\langle t_{zi}^\dagger \rangle = \langle t_{zi} \rangle = t$, and impose the constraint (6.22) at a mean-field level by introducing a Lagrange multiplier μ as was done in the SU(2) case. I will use this approach in the following since, as shown by Muniz *et al* [9], it gives a better result than the former compared with data from Monte Carlo simulations. I add a magnetic field term $-h\sum_r S_r^z$ to (6.18) and write the Hamiltonian as

$$\begin{aligned}
H = \frac{J}{2}\sum_{r,a} t^2[&(d_r^\dagger d_{r+a} + u_r^\dagger u_{r+a} + u_r d_{r+a} + d_r^\dagger u_{r+a}^\dagger + \text{H.c.})\\
&+ (u_r^\dagger u_r - d_r^\dagger d)(u_{r+a}^\dagger u_{r+a} - d_{r+a}^\dagger d_{r+a})] + D\sum_r (u_r^\dagger u_r + d_r^\dagger d_r)\\
&- h\sum_r (u_r^\dagger u_r - d_r^\dagger d_r) - \sum_r \mu_r (u_r^\dagger u_r + d_r^\dagger d_r + t^2 - 1).
\end{aligned} \tag{6.31}$$

A temperature-dependent Lagrange multiplier μ_r is introduced to impose the constraint of single-site occupancy, as mentioned above. In a mean-field approximation, we replace the local constraint by a global one, i.e. $\mu_r = \mu$. In the next step, I make the mean-field decoupling

$$\begin{aligned}(u_r^\dagger u_r - d_r^\dagger d_r)(u_{r+\delta}^\dagger u_{r+\delta} - d_{r+\delta}^\dagger d_{r+\delta}) = &\frac{1}{2}(1 - t^2 + m)\\ &\times [(u_r^\dagger u_r + u_{r+\delta}^\dagger u_{r+\delta})\\ &+ (d_r^\dagger d_r + d_{r+\delta}^\dagger d_{r+\delta})]\\ &- p(u_r d_{r+\delta} + d_r u_{r+\delta} + \text{H.c.})\\ &- \frac{1}{2}(1 - t^2) - \frac{m^2}{2} + 2p^2,\end{aligned} \tag{6.32}$$

where $\langle d_r^\dagger u_{r+\delta}^\dagger \rangle = \langle d_r u_{r+\delta} \rangle = p$, and $m = \langle u_r^\dagger u_r \rangle - \langle d_r^\dagger d_r \rangle$. In the absence of an external magnetic field, one has $m = 0$.

Taking the Fourier transform and performing a Bogoliubov transformation given by

$$u_k^\dagger = \tilde{u}_k \alpha_k - v_k \beta_k, \qquad d_{-k} = -v_k \alpha_k + \tilde{u}_k \beta_k, \tag{6.33}$$

where

$$\tilde{u}_k = \frac{1}{\sqrt{2\omega_k}}(\Lambda_k + \omega_k)^{1/2}, \qquad v_k = \frac{1}{\sqrt{2\omega_k}}(\Lambda_k - \omega_k)^{1/2}, \tag{6.34}$$

we obtain

$$H = \sum_k \left(\omega_k^{(1)} \alpha_k^\dagger \alpha_k + \omega_k^{(2)} \beta_k^\dagger \beta_k\right) + \sum_k (\omega_k - \Lambda_k) + C, \tag{6.35}$$

with

$$\omega_k^{(1)} = \omega_k - h + 2m, \qquad \omega_k^{(2)} = \omega_k + h - 2m$$

$$\omega_k = \sqrt{\Lambda_k^2 - \Delta_k^2}, \quad \Lambda_k = -\mu + D + 2(1 - t^2) + 4t^2\gamma_k, \quad \Delta_k = 4(t^2 - p)\gamma_k,$$

where

$$\begin{aligned}\gamma_k &= \frac{1}{z}\sum_a e^{i\vec{k}.\vec{a}}.\\ C &= \mu N(1 - t^2) - N(1 - t^2)^2 + 4Np^2.\end{aligned} \tag{6.36}$$

In the following, I will consider the case $h = 0$. In the mean-field approximation, the free energy is given by:

$$G = NE_g - \frac{2}{\beta}\sum_k \ln[1 + n(k)], \tag{6.37}$$

where $n(k) = 1/(e^{\beta\omega_k} - 1)$, and E_g is the ground state energy per site:

$$E_g = \frac{1}{N}\sum_k(\omega_k - \Lambda_k) + C/N. \tag{6.38}$$

Minimizing G with relation to μ and t^2, we obtain the following self-consistent equations, which should be solved numerically:

$$p = -\frac{1}{2\pi^2}\int_0^\pi \int_0^\pi d\vec{k}\frac{\gamma_k \Delta_k}{\omega_k}\coth\left(\frac{\beta\omega_k}{2}\right), \tag{6.39}$$

$$\mu = \frac{4}{\pi^2}\int_0^\pi \int_0^\pi d\vec{k}\frac{\gamma_k(\Lambda_k - \Delta_k)}{\omega_k}\coth\left(\frac{\beta\omega_k}{2}\right), \tag{6.40}$$

$$2 - t^2 = \frac{1}{\pi^2}\int_0^\pi \int_0^\pi d\vec{k}\frac{\Lambda_k}{\omega_k}\coth\left(\frac{\beta\omega_k}{2}\right). \tag{6.41}$$

The transition from the large D phase to the Néel phase happens when the energy gap goes to zero, and this condition determines D_C. Note that in the ordered phase, there is a Goldstone mode associated with the spontaneous breaking of the U(1) symmetry of global spin rotations along the z-axis. A numerical calculation [5] yields $D_C/J = 5.47$. The details of the above calculations can be found in [5, 6]. The ordered Neél phase for $D < D_C$ and $T = 0$ was studied by Zhang [10], and an explicit calculation was given by Pires [11].

It should be noted that, as was mentioned in section 2.5, for D smaller than D_C a Berezinskii–Kosterlitz–Thouless (BKT) phase transition occurs at a critical temperature T_{BKT}. Therefore, the above formalism can be applied only at zero temperature in this region. The phase diagram is shown in figure 6.1. The BKT phase occurs to the left $D < D_C$ of the critical point D_C. The solid line represents the T_{BKT} line of critical points determined by the BKT transition. The dashed line represents a crossover from the quantum critical region to the phase-disordered regime, where quantum fluctuations dominate [12].

Now I briefly discuss what happens when the external magnetic field h is present. The lower component of the dispersion relation decreases with the increasing of h, and at a critical magnetic field h_{c1}, the energy gap vanishes. We assume that the energy gap remains zero when the magnetic field is further increased, and part of the excitations (the α magnons) condense. A magnetization parallel to the external field appears, and at the same time, a staggered magnetization in the xy-plane (let's say in the x-direction) occurs. At a second critical magnetic field h_{c2}, the magnetization saturates, and the staggered magnetization vanishes. For $h_{c1} < h < h_{c2}$, a thermodynamic phase transition at a critical temperature $T_c(h)$ takes place. This field-induced antiferromagnetic transition corresponds to a Bose–Einstein condensation (the applied magnetic field plays the role of a chemical potential). In the region, $h_{c1} < h < h_{c2}$, the bosons form a Bose–Einstein condensate that corresponds to the XY antiferromagnetic ordering in the original spin language [13].

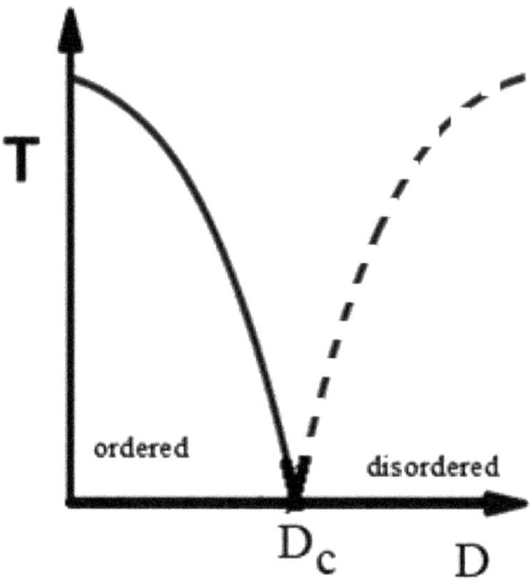

Figure 6.1. Phase diagram of Hamiltonian (6.16).

6.4 Bilinear biquadratic model

Another model that presents quadrupolar phases and has been intensively studied in the literature is the spin-1 bilinear biquadratic model described by the following Hamiltonian

$$H = J\sum_{\langle n,m\rangle} \vec{S}_n \cdot \vec{S}_m + K\sum_{\langle n,m\rangle} (\vec{S}_n \cdot \vec{S}_m)^2, \tag{6.42}$$

where K is the coupling for the biquadratic interaction between nearest-neighbor spin pairs. In this section, I will study this model on the square lattice. It is usual in the literature to adopt the notation

$$J = J_1 \cos\theta, \qquad K = J_1 \sin\theta. \tag{6.43}$$

This model was first studied by Papanicolaou [7] using a variational approach, and a recent study by Niesen and Corboz [15] using a tensor network algorithm has shown that the model presents the following phases:

(i) An antiferromagnetic Néel phase for $0.189\pi > \theta > -\pi/2$.
(ii) A ferroquadrupolar phase for $-\pi/2 > \theta > -3\pi/4$.
(iii) A ferromagnetic phase for $5\pi/4 < \theta < \pi/2$.
(iv) The remaining region is more complex, presenting several phases, such as a three-sublattices antiferroquadrupolar phase (with perpendicular directors on neighboring sites) and a three-sublattice 120° magnetically ordered phase.

A quadrupolar sate is described uniquely by the vector $\vec{Q}$, and its magnetic dipole moment m is zero. The $|0\rangle$ state in the S^z basis is an example of this state. The state $|0\rangle$ breaks spin-rotation symmetry, even in the absence of magnetization since $\langle (S^z)^2\rangle = 0$. The fluctuations in x and y-directions, on the other side, are nonzero.

That is: $\langle (S^x)^2 \rangle = \langle (S^y)^2 \rangle = 1$. We can think of this state, in the classical limit, as a dipole moment that fluctuates in the x-y plane such that it is zero on average. This state can be represented as a disc in the plane of fluctuations. An important object is the *director*, which is a vector perpendicular to the plane of fluctuations. For the state $|0\rangle$, the director is $\pm\hat{e}$. The orientation of the director describes completely any quadrupolar single-particle state. Using the basis (6.20), we can expand any arbitrarily spin one single-particle state as $|\vec{d}_j\rangle = d_j^x|x\rangle + d_j^y|y\rangle + d_j^z|z\rangle$, where d_j^x, d_j^y and d_j^z are complex numbers. The vector $\vec{d} = (d_x, d_y, d_z)$ can be written as $\vec{d} = \vec{u} + i\vec{v}$ [16], where $\vec{u}$ and $\vec{v}$ are reals, and so we can write the magnetic moment of $\vec{d}$ as $\vec{S} = 2\vec{u} \times \vec{v}$. This state can be normalized using global phase invariance. That allows us to write $\vec{u}_i^2 + \vec{v}_i^2 = 1$ and $\vec{u}_i \cdot \vec{v}_i = 0$. Note that $\vec{u} = \vec{v}$, describes fully magnetized states, and $(u, v) = (0, 1)$ or $(1, 0)$, describes non-magnetic states ($m = 0$, which are purely quadrupolar states).

When studying quadrupolar order, it is more convenient to use the time-reversal invariant basis (6.21). As an example, I will examine the ferroquadrupolar phase following Penc and Laucchli [17]. The quadrupolar solution corresponds to the condensation of the appropriate boson. All states defined by (6.20) have $\langle S^z \rangle = 0$. Hence, we can condense any t_α boson, which we chose as the t_y operator, and write

$$t_{yi}^\dagger,\, t_{yi} \to \sqrt{1 - t_{xi}^\dagger t_{xi} - t_{zi}^\dagger t_{zi}}\,. \tag{6.44}$$

Spin and quadrupolar excitations are captured by the remaining two flavors. We can write Hamiltonian (6.42) as

$$H = (J - K)\sum_{\langle ij \rangle} \vec{S}_i \cdot \vec{S}_j + K\sum_{\langle ij \rangle} \vec{A}_i \cdot \vec{A}_j, \tag{6.45}$$

where

$$\vec{A}_i \cdot \vec{A}_j = \sum_{\mu\nu} t_{\mu i}^\dagger t_{\nu i} t_{\nu j}^\dagger t_{\mu j}. \tag{6.46}$$

Expanding (6.44), keeping up to bilinear terms, substituting into (6.45), performing a Fourier transform, and neglecting a constant term, we obtain

$$H = 2\sum_{\nu=x,z}\sum_k [2A_k t_{\nu k}^\dagger t_{\nu k} + B_k(t_{\nu k} t_{\nu -k} + t_{\nu k}^\dagger t_{\nu -k}^\dagger)], \tag{6.47}$$

where

$$A_k = J\gamma_k - K,\ \ B_k = (K - J)\gamma_k. \tag{6.48}$$

Using the Bogoliubov transformation

$$t_{\nu,k} = u_k \alpha_{\nu,k} + v_k \alpha_{\nu,\,-k}^\dagger, \tag{6.49}$$

to diagonalize (6.47), we get

$$H = \sum_{\nu=x,z} \sum_k \left[\omega_k \alpha^\dagger_{\nu,k} \alpha_{\nu,k} + \left(\frac{\omega_k}{2} - A_k \right) \right], \tag{6.50}$$

where

$$u_k^2 = \frac{A_k + \omega_k}{2\omega_k}, \quad v_k^2 = \frac{A_k - \omega_k}{2\omega_k}, \quad \omega_k = 2\sqrt{A_k^2 - B_k^2}. \tag{6.51}$$

The dispersion is degenerate in the flavor index $\nu = x, y$. Since the ferroquadrupolar phase spontaneously breaks the spin-rotational symmetry, there are two gapless Goldstone modes at $\vec{k} = 0$. From equation (6.23) we have

$$S_i^z = i\left(t^\dagger_{yi} t_{xi} - t^\dagger_{xi} t_{yi}\right). \tag{6.52}$$

Using (6.44) we can write

$$S^{zz}(\vec{k}, t) = \langle S_k^z S_{-k}^z(t) \rangle = -\left\langle [t_{x,k} - t^\dagger_{x,k}][t_{x,-k}(t) - t^\dagger_{x,-k}(t)] \right\rangle. \tag{6.53}$$

Using now (6.49) and (6.51) and taking the Fourier transform, we get at zero temperature

$$S^{zz}(\vec{k}, \omega) = \sqrt{\frac{A_k + B_k}{A_k - B_k}}\, \delta(\omega - \omega_k). \tag{6.54}$$

We have $S^{xx}(\vec{k}, \omega) = S^{zz}(\vec{k}, \omega)$. There is no Bragg peak as the dynamical spin structure factor has a vanishing spectral weight at $\vec{k} = 0$, $\omega = 0$, because of the conservation of time-reversal symmetry in quadrupolar states. The calculation of $S^{yy}(\vec{k}, \omega)$ involves four operators, and as a consequence, we see that nematic waves are bound states of two magnons. The quadrupolar structure factor should exhibit Bragg peaks at the ordering vector $\vec{k} = 0$.

For a recent study of the antiferroquadrupolar phase, see Hu *et al* [17]. For magnetic order, following the usual procedure of spin-wave theory, we introduce local rotations to align spins at different sites, along the $|1\rangle$ state. The reader can easily perform the calculations for the ferromagnetic and antiferromagnetic phases. For the ferromagnetic phase, the dispersion relations are

$$\omega_{k1} = 4J(\gamma_k - 1), \quad \omega_{k2} = 4[K\gamma_k + (K - 2J)]. \tag{6.55}$$

While for the antiferromagnetic phase, we have

$$\omega_{ki} = \sqrt{A_{ki}^2 - B_{ki}^2}, \tag{6.56}$$

with

$$A_{k1} = 4(J - K), \quad B_{k1} = 4(J - K)\gamma_k, \quad A_{k2} = 4(2J - K), \quad B_{k2} = 4K\gamma_k. \tag{6.57}$$

To illustrate the technique, I will consider the following example studied by Liu *et al* [18], described by the spin -1 Hamiltonian:

$$H = \frac{1}{2}\left\{\sum_{\langle ij\rangle}[J_1\vec{S}_i \cdot \vec{S}_j - K_1(\vec{S}_i \cdot \vec{S}_j)^2] + \sum_{\ll ij\gg}[J_2\vec{S}_i \cdot \vec{S}_j - K_2(\vec{S}_i \cdot \vec{S}_j)^2]\right\}, \tag{6.58}$$

where J_1 and K_1 are exchange coupling for the nearest-neighbor bonds on a square lattice, and J_2 and K_2 are for the next nearest-neighbor bonds. It is believed that this Hamiltonian can be used to treat a $(\pi, 0)$ antiferromagnetic order in the parent 122 iron pnictides, such as $BaFe_2As_2$ [19]. The ground state that I will be interested in is a $\vec{Q} = (\pi, 0)$ collinear antiferromagnetic order. It is assumed that the corresponding classical spin configuration of this state has all spins aligned along the S^z-direction. To proceed, we perform a site-dependent spin rotation about the y-axis and get

$$\vec{\tilde{S}}_i = \hat{R}_y(\vec{Q} \cdot \vec{r}_i)\vec{S}_i. \tag{6.59}$$

After this rotation, the spins in even columns remain unchanged, while those in odd columns are rotated by a π angle around the y-axis. Hence, in the rotating frame, all spins align ferromagnetically along the negative $\tilde{S}_z$-direction. Now we have magnetic order, and therefore we use the operators u and d. The magnetic ground state is now polarized, and thus this state corresponds to a condensation of the d boson on each site. Using the constraint $u^\dagger u + d^\dagger d + t_z^\dagger t_z = 1$ we have

$$d_i \approx \sqrt{1 - u_i^\dagger u_i - t_{zi}^\dagger t_{zi}}\,. \tag{6.60}$$

Using (6.27) and expanding the Hamiltonian in terms of u_i and t_{zi} up to the quadratic order, we get

$$\begin{aligned} H_1 = \frac{1}{2}\sum_K [&A_{ko}(t_{kz}^\dagger t_{kz} + t_{-kz}^\dagger t_{-kz}) + A_{k1}(u_k^\dagger u_k + u_{-k}^\dagger u_{-k}) \\ &+ B_{k0}(t_{kz}^\dagger t_{-kz}^\dagger + t_{kz}t_{-kz}) + B_{k1}(u_k^\dagger u_{-k} + u_k u_{-k})], \end{aligned} \tag{6.61}$$

where

$$\begin{aligned} A_{k0} &= 2J_1 \cos k_y + 2K_1 + 4(J_2 + K_2), \\ B_{k0} &= -2(J_1 + K_1)\cos k_x - 4(J_2 + K_2)\cos k_x \cos k_y, \\ A_{k1} &= 8J_2 - 2K_1 \cos k_y + 4K_2, \\ B_{k1} &= -2K_1 \cos k_x - 4K_2 \cos k_x \cos k_y. \end{aligned} \tag{6.62}$$

We diagonalize (6.61) using the Bogoliubov transformation

$$u_k = \tilde{u}_{k1}\beta_{k1} + v_{k1}\beta_{-k1}^\dagger, \qquad t_{kz} = \tilde{u}_{k0}\beta_{k0} + v_{k0}\beta_{-k0}^\dagger, \tag{6.63}$$

where

$$\tilde{u}_{ki} = \sqrt{\frac{A_{ki} + \omega_{ki}}{2\omega_{ki}}}, \quad v_{ki} = -\mathrm{sgn}(B_{ki})\sqrt{\frac{A_{ki} - \omega_{ki}}{2\omega_{ki}}}, \quad \omega_{ki} = \sqrt{A_{ki}^2 - B_{ki}^2}, \quad (6.64)$$

and $i = 0, 1$, and get

$$H_1 = \sum_k \sum_{i=0,1} \omega_{ki}\beta_{ki}^\dagger \beta_{ki} + \text{constant}. \quad (6.65)$$

The operator $u^\dagger$ creates an excitation with $\Delta S^z = 2$ corresponding to creating a spin-2 quadrupolar excitation, while $t_z^\dagger$ creates an excitation with $\Delta S^z = 1$. ω_{k0} is the conventional magnon mode (Goldstone mode) with gapless excitation, ω_{k1} has a quadrupolar character, which is the bound state of two magnons. (For more details, see ref. [18].)

A final remark: For spins $S > ½$, multipolar ordering (where $\langle \vec{S} \rangle = 0$) of the type $\langle S_{\alpha 1} S_{\alpha 2} \ldots S_{\alpha n} \rangle$, with $n = 1$ (dipolar), $n = 2$ (quadrupolar), $n = 3$ (octupolar), etc, up to $n = 2S$ can exist. In all these cases, the ground state order does not have a classical counterpart at $T = 0$.

We can also use the SU(3) formalism in a mean-field approach similar to the one we did for the SU(2) case in chapter 5. As was said before, unlike the spin-wave theory, the mean-field method does not prescribe an ordered state in advance. We write the Hamiltonian in terms of all the t operators to get a quartic Hamiltonian. We then decouple the four-operator terms into a bilinear form and use a mean-field ansatz to get the solutions [18]. When the spectrum becomes gapless, a Bose condensation occurs, signaling the onset of an ordered phase.

6.5 Variational approach

To use the SU(3) spin-wave formalism (as has been done above), we should know, at least to start the calculation, the kind of order we have in each region when we vary the system's parameters. We can then check if our original assumption was correct. A simple procedure that gives us an idea of the phases involved is a variational theory briefly described here. Using the relation

$$Q^{\alpha\beta} = S^\alpha S^\beta + S^\beta S^\alpha - \frac{4}{3}\delta_{\alpha\beta}, \ (\alpha, \beta = x, y, z) \quad (6.66)$$

we can write Hamiltonian (6.42) as

$$H = \sum_{n,m}\left(J - \frac{K}{2}\right)\vec{S}_n \cdot \vec{S}_m + \frac{1}{2}\sum_{n,m} K\left(\vec{Q}_n \cdot \vec{Q}_m + \frac{8}{3}\right). \quad (6.67)$$

The term $\vec{S}_n \cdot \vec{S}_m$ is maximal when the two spins are parallel and minimal when they are antiparallel. A pure quadrupolar state (when both $\vec{d}_n$ and $\vec{d}_m$ are real) has the expectation value

$$\langle \vec{Q}_n \cdot \vec{Q}_m \rangle = |\vec{d}_n \cdot \vec{d}_m|^2 + |\vec{d}_n^{\,*} \cdot \vec{d}_m|^2 - \frac{2}{3} \, (n \neq m), \tag{6.68}$$

which is maximal for collinear directors ($\vec{d}_n \parallel \vec{d}_m$) and minimal for perpendicular ones ($\vec{d}_n \perp \vec{d}_m$). The term $\langle \vec{Q}_n \cdot \vec{Q}_m \rangle$ is minimized if $\vec{d}_n$ is orthogonal both to $\vec{d}_m$ and $\vec{d}_m^{\,*}$, which implies that one state is a pure quadrupole with director $\vec{d}$, while the other one may feature either a pure quadrupole with its director perpendicular to $\vec{d}$ or a spin vector of arbitrary length pointing along $\vec{d}$. Papanicolaou [7] called this phase 'semi-ordered.'

The variational wave function is taken as the direct product of one-body wave function as [20]

$$|\psi\rangle = \prod_n |\psi_n\rangle, \quad |\psi_n\rangle = d_{nx}\,|x\rangle + d_{ny}\,|y\rangle + d_{nz}\,|z\rangle, \tag{6.69}$$

where $d_{n\alpha}$ is a complex coefficient satisfying the normalization condition $|\vec{d}_n| = 1$. The expectation values of the spin and the quadrupole operators have simple expressions in these states as follows:

$$\begin{aligned} \langle \vec{d}_n | \vec{S} | \vec{d}_n \rangle \cdot \langle \vec{d}_m | \vec{S} | \vec{d}_m \rangle &= |\langle \vec{d}_n | \vec{d}_m \rangle|^2 - |\langle \vec{d}_n | \vec{d}_m^{\,*} \rangle|^2, \\ \langle \vec{d}_n | \vec{Q} | \vec{d}_n \rangle \cdot \langle \vec{d}_m | \vec{Q} | \vec{d}_m \rangle &= |\langle \vec{d}_n | \vec{d}_m \rangle|^2 + |\langle \vec{d}_n | \vec{d}_m^{\,*} \rangle|^2 - \frac{2}{3}. \end{aligned} \tag{6.70}$$

The energy evaluated using the variational wave function is given as

$$E_g = \langle \psi | H | \psi \rangle = \sum_{n,m} \left[J\,|\vec{d}_n^{\,*} \cdot \vec{d}_m|^2 + (K - J)|\vec{d}_n \cdot \vec{d}_m|^2 \right] + \text{const.} \tag{6.71}$$

If the ordering vector is $\vec{q} = 0$, the wave functions on each of the sublattices are equal, $\vec{d}_n = \vec{d}_m = \vec{d}$. The energy per site is then given by

$$E_g = 4K + 8(J - K)u^2 v^2 = 4K + 8(J - K)u^2(1 - u^2), \tag{6.72}$$

where I have used the condition $u^2 + v^2 = 1$. For J—$K < 0$ ($\pi/2 < \theta < 5\pi/4$) the minimum is given by $u = v = 1/\sqrt{2}$, which corresponds to the ferromagnetic state with energy $E_g = 2(K + J)$. If $J - K > 0$ ($-3\pi/4 < \theta < -\pi/2$) we have two solutions for the minimum. The first is $u = 1$, $v = 0$, therefore $\vec{d}$ is real. This phase is the ferroquadrupolar phase, whose director is oriented in the arbitrary direction $\vec{u}$. The other solution is $u = 0$, $v = 1$, with director $\vec{v}$. Both solutions have energy $E_g = 4K$. In the ferroquadrupolar phase, directors are parallel to each other but may point anywhere.

The variational theory gives a two-sublattice Néel antiferromagnet for $-\pi/2 \leqslant \theta \leqslant \pi/4$ [7]. The region $\pi/4 \leqslant \theta \leqslant \pi/2$ is more complicated with a highly degenerate ground state. We can compare the phase diagram obtained using the variational procedure with the numerical calculations performed by Niesen and Corboz [14].

If we add a magnetic field to the system, we have a term $-h\sum_i S_i^z$ that becomes

$$ih \sum_i \left(\vec{d}_i^* \times \vec{d}_i\right)^z. \tag{6.73}$$

In the ground state, the magnetic field confines the $\vec{d}$-vector within the xy-plane. That happens because h couples to $d_x^* d_y - d_y^* d_z$ but not to d_z. When $\vec{d}$ is real, a pure quadrupole is formed symmetrically about the director $\vec{d}$. At $h \neq 0$ the emergent imaginary component of $\vec{d}$ will distort it by shifting its fluctuation center toward the $+z$-direction. In conclusion, the director turns perpendicular to the magnetic field, and the quadrupole starts developing a dipole moment parallel to the magnetic field [21]. The energy (6.72) in the presence of the magnetic field becomes

$$E = 4K + 8(J - K)u^2v^2 - 2h(u_x v_y - u_y v_x). \tag{6.74}$$

In terms of the local magnetization $m = 2uv$, the energy can be written as $E = 4K + 2(J - K)m^2 - hm$. Minimizing the energy, we find that the magnetization grows linearly with the magnetic field as $m = h/[4(J - K)]$. The ferroquadrupolar state finally turns into a fully polarized magnet at $h = 4(J—K)$. We can write $m = \sin\mu$, and then $\vec{d} = (\cos(\mu/2), i\sin(\mu/2), 0)$, in the ferroquadrupolar phase. To use the spin-wave formalism in the ferroquadrupolar phase we can perform a unitary transformation V_i according to the directors in the magnetic field to rotate the t_α operators. The transformation matrix V_i is identical on all sites i in the FQ state [20]:

$$V_i = \begin{pmatrix} i\sin(\mu/2) & 0 & \cos(\mu/2) \\ \cos(\mu/2) & 0 & i\sin(\mu/2) \\ 0 & 1 & 0 \end{pmatrix}. \tag{6.75}$$

Hu *et al* [17] (see also [20]) have presented the phase diagram, using a variational procedure, for the biquadratic model on the square lattice, including the first three neighbors. In this case, equation (6.71) becomes

$$E_g = \sum_{n,m}\left[J_{nm}\,|\vec{d}_n^* \cdot \vec{d}_m|^2 + (K_{nm} - J_m)|\vec{d}_n \cdot \vec{d}_m|^2 + K_{nm}\right]. \tag{6.76}$$

6.6 Exercises

6.1. Prove equation (6.33).
6.2. Prove equation (6.52).
6.3. Prove equations (6.53) and (6.54).
6.4. Prove equation (6.62).

References

[1] Sachdev S and Bhatt R N 1990 Bond-operator representation of quantum spins: mean-field theory of frustrated quantum Heisenberg model *Phys. Rev.* B **41** 9323–9

[2] Matshushita Y, Gelgand M P and Ishii C 1999 Bond-operator mean-field theory for the bilayer Heisenberg model *J. Phys. Soc. Jpn.* **68** 247–52

[3] Shik H Y, Li Q Y and Lin H Q 2003 Constructing soluble quantum spin models *Nucl. Phys.* B **666** 337–60

[4] Sachdev S 2011 *Quantum Phase Transitions* (Cambridge: Cambridge University Press)

[5] Wang H T and Wang Y 2005 Long-range order in gapped magnetic systems induced by Bose–Einstein condensation *Phys. Rev.* B **71** 104429

[6] Pires A S T 2011 The spin-1 anisotropic Heisenberg antiferromagnet on a square lattice at low temperatures *J. Magn. Magn. Mater.* **323** 1977–9

[7] Papanicolaou N 1988 Unusual phases in quantum spin-1 systems *Nucl. Phys.* B **305** 367–95

[8] Arovas D P and Auerbach A 1988 Functional integral theories of low-dimensional quantum Heisenberg models *Phys. Rev.* B **38** 316–32

[9] Muniz R A, Kato Y and Batista C D 2014 Generalized spin-wave theory: application to the bilinear-biquadratic model *Prog. Theor. Exp. Phys.* **2014** 083101

[10] Zhang Z, Wierschem K, Yap I, Kato Y, Batista C D and Sengupta P 2013 Phase diagram and magnetic excitations of anisotropic spin-one magnets *Phys. Rev.* B **87** 174405

[11] Pires A S T 2018 Effect of the Higgs mode in the spin transport of a Heisenberg antiferromagnet on a square lattice *Solid State Commun.* **281** 22–6

[12] Philips P 2003 *Advanced Solid State Physics* (Boulder, CO: Westview Press)

[13] Zapf V S, Zocco D, Hansen B R, Jaime M, Harrison N, Batista C D, Kenzelmann M, Niedermayer C, Lacerda A and Paduan-Filho A 2006 Bose–Einstein condensation of $S = 1$ Nickel spin degrees of freedom in $NiCl_2$–$4SC(NH_2)_2$ *Phys. Rev. Lett.* **96** 077204

[14] Niesen I and Corboz P 2017 A tensor network study of the complete ground-state phase diagram of the spin-1 bilinear-biquadratic Heisenberg model on the square lattice *SciPost Phys.* **3** 030

[15] Hu W J, Lai H H, Gong S S, Yu R, Dagotto E and Si Q 2020 Quantum transitions of nematic phases in a spin-1 bilinear biquadratic model and their implication for FeSe *Phys. Rev. Res.* **2** 023359

[16] Penc K and Lauchli A M 2011 Spin nematic phases in quantum spin systems *Introduction to Frustrated Magnetism* ed C Lacroix, P Mendels and F Mila (Berlin: Springer) pp 331–62

[17] Hu W J, Gong S S, Lai H H, Hu H, Si Q and Nevidomsky H 2019 Nematic spin liquid phase in a frustrated spin-1 system on the square lattice *Phys. Rev.* B **100** 165142

[18] Li P, Zhang G M and Shen S Q 2007 The SU(3) bosons and the spin nematic state of the spin-1 bilinear-biquadratic triangular lattice *Phys. Rev.* B **75** 104420

[19] Liu C, Lu X, Dai P, Yu R and Si Q 2020 Anisotropic magnetic excitations of a frustrated bilinear-biquadratic spin model: implications for spin waves of detwinned iron pnictides *Phys. Rev.* B **101** 024510

[20] Wang Z, Hu W J and Nevidomsky A H 2016 Spin ferroquadrupolar order in the nematic phase of FeSe *Phys. Rev. Lett.* **116** 247203

[21] Tanaka K and Hotta C 2020 Finite temperature thermodynamic properties of the spin-1 nematics in an applied magnetic field *Phys. Rev.* B **102** 140401

IOP Publishing

Theoretical Tools for Spin Models in Magnetic Systems

Antonio Sergio Teixeira Pires

Chapter 7

Dynamics

This chapter introduces the linear response theory, which provides a general framework for studying the dynamical behavior of a many-body system close to thermal equilibrium. The dynamical processes are either due to external perturbations or spontaneous fluctuations, and these two processes are interrelated.

7.1 Linear response theory

In an experiment, a localized disturbance is created by an external field in the neighborhood of some point $\vec{r}$ at some time t. Then the response of the system is measured at some other point $\vec{r}\,'$ at a later time $t' > t$. I will leave the spatial dependence aside for the moment.

To start, we consider an isolated system described by the Hamiltonian H_0, add an interaction term with an external field, and write

$$H = H_0 + V(t). \tag{7.1}$$

We confine ourselves with weak perturbation and study the response of the system in the linear approximation. In this approximation, the density matrix of the system is given by

$$\rho(t) = \rho_0 + \Delta\rho(t), \tag{7.2}$$

where ρ_0 is the density matrix of the unperturbed system, given by $\rho_0 = e^{-\beta H_0}/Z$, where $Z = tre^{-\beta H_0}$. We suppose that the external field is gradually turned on starting at $t = -\infty$ and write $V(t)e^{\eta t}$ where η is a small positive quantity. I will not explicitly write the term $e^{\eta t}$, but we should be aware that it exists and put it back wherever necessary. Any unwanted secondary effects may be eliminated by taking the limit $\eta \to 0^+$. We have then $\rho(t = -\infty) = \rho_0$. We also suppose that the system

doi:10.1088/978-0-7503-3879-0ch7

is in thermal equilibrium before the external field is applied. The equation of motion for ρ is

$$i\frac{\partial\rho(t)}{\partial t} = [H, \rho] = [H_0, \rho(t)] + [V(t), \rho(t)], \tag{7.3}$$

where I have set $\hbar = 1$. In the linear response theory, we neglect the term $V\Delta\rho$ and write

$$i\Delta\dot{\rho}(t) = [H_0, \Delta\rho(t)] + [V(t), \rho_0]. \tag{7.4}$$

Equation (7.4) can be written in the form

$$\Delta\rho(t) = -i\int_{-\infty}^{t} e^{-iH_0(t-t')}[V(t'), \rho_0]e^{iH_0(t-t')}dt', \tag{7.5}$$

as can be seen by calculating $\Delta\dot{\rho}(t)$.

For any operator A, the following identity may be verified by direct differentiation with respect to an auxiliary variable λ:

$$\frac{d}{d\lambda}\left\{e^{\lambda H_0}[A, e^{-\lambda H_0}]\right\} = H_0 e^{\lambda H_0}Ae^{-\lambda H_0} - e^{\lambda H_0}Ae^{-\lambda H_0}H_0 = [H_0, A(-i\lambda)], \tag{7.6}$$

where

$$A(-i\lambda) \equiv e^{\lambda H_0}Ae^{-\lambda H_0}. \tag{7.7}$$

Considering $i\lambda$ as an imaginary time, we write

$$[H_0, A(-i\lambda)] \equiv -i\dot{A}(-i\lambda). \tag{7.8}$$

Note that while in equation (7.3) we have H, all thermal averages are calculated using H_0 in the linear response theory. Integrating (7.6) from 0 to β we obtain

$$[A, \rho_0] = -i\rho_0\int_0^{\beta} \dot{A}(-i\lambda)d\lambda. \tag{7.9}$$

Using equations (7.8) and (7.9) we can write equation (7.5) as

$$\Delta\rho(t) = -i\int_{-\infty}^{t} e^{-iH_0(t-t')}\left[-i\rho_0\int_0^{\beta} \dot{V}(-i\lambda)d\lambda\right]e^{iH_0(t-t')}dt'. \tag{7.10}$$

Writing $\tau = t - t'$ and then $\tau \to t$ we get

$$\Delta\rho = -\rho_0\int_0^{\infty} dt\int_0^{\beta} d\lambda\dot{V}(-t - i\lambda). \tag{7.11}$$

The average of an operator O at time t is given by

$$\langle O(t)\rangle = tr[\rho(t)O], \quad \text{with } tr\,\rho(t) = 1. \tag{7.12}$$

The induced change in the variable B from its equilibrium value is

$$\Delta B(t) = tr(\rho B) = tr[(\rho_0 + \Delta\rho)B] = \langle B\rangle_0 + tr(\Delta\rho B), \tag{7.13}$$

where $\langle B\rangle_0 = tr(\rho_0 B)$. The linear response theory is restricted to the regime where ΔB changes linearly with the field. I will take $\langle B\rangle_0 = 0$. For example, B could be the magnetization and $V(t)$ a time-dependent applied magnetic field. Writing the perturbation as $V = -Af(t)$, where A is a constant operator, and using equation (7.5) we find

$$\Delta B(t) = itr\int_{-\infty}^{t} e^{-iH_0(t-t')}[A, \rho_0]e^{iH_0(t-t')}Bf(t')dt'. \tag{7.14}$$

Using $tr(AB) = tr(BA)$ we have

$$\Delta B(t) = itr\int_{-\infty}^{t} [A, \rho_0]B(t - t')f(t')dt', \tag{7.15}$$

where $B(t) = e^{iH_0t}Be^{-iH_0t}$. We can write equation (7.15) as

$$\Delta B(t) = i\int_{-\infty}^{t} \langle[B(t - t'), A]\rangle_0 f(t')dt'. \tag{7.16}$$

Defining

$$K_{BA}(t) \equiv i\langle[B(t), A]\rangle_0, \tag{7.17}$$

equation (7.16) takes the form

$$\Delta B(t) = \int_{-\infty}^{t} K_{BA}(t - t')f(t')dt'. \tag{7.18}$$

Note that $K_{BA}(t) = -K_{AB}(-t)$. If $f(t)$ is of the form $f(t) = e^{-i\omega t}f$, we define the dynamical susceptibility $\chi(\omega)$ as

$$\Delta B(t) = \chi(\omega)f(t). \tag{7.19}$$

Then

$$\chi_{BA}(\omega) = \int_{-\infty}^{t} e^{-i\omega t'}K_{BA}(t - t')dt'. \tag{7.20}$$

Changing $t - t' = \tau$ and then taking $\tau = t$ we get

$$\chi_{BA}(\omega) = \int_{0}^{\infty} dte^{i\omega t}K_{BA}(t). \tag{7.21}$$

Using equation (7.17), we write (7.21) as

$$\chi_{BA}(\omega) = itr\int_{0}^{\infty} [A, \rho_0]B(t)e^{i\omega t}dt. \tag{7.22}$$

Using equation (7.9), we get

$$\chi_{BA}(\omega) = \int_0^\infty e^{i\omega t}dt\left[tr\int_0^\beta \rho_0\dot{A}(-i\lambda)B(t)d\lambda\right]$$
$$= -\int_0^\infty dte^{i\omega t}tr\left[\rho_0\int_0^\beta A(-i\lambda)\frac{dB(t)}{dt}d\lambda\right], \tag{7.23}$$

or

$$\chi_{BA}(\omega) = -tr\ \rho_0\int_0^\beta A(-i\lambda)\left(\int_0^\infty e^{i\omega t}\frac{dB}{dt}dt\right)d\lambda, \tag{7.24}$$

where I have used equation (7.8) and the relation $tr(\rho_0 H_0 AB) = tr(AB\rho_0 H_0)= tr(ABH_0\rho_0) = tr(\rho_0 ABH_0)$. Integrating by parts the above equation, we find

$$\chi_{BA}(\omega) = tr\ \rho_0\int_0^\beta A(-i\lambda)\left[B(0) - i\omega\int_0^\infty e^{i\omega t}B(t)dt\right]d\lambda. \tag{7.25}$$

We define the *relaxation function* $R_{AB}(t)$ of two arbitrary operators A and B by

$$R_{AB}(t) \equiv (A,\ B(t)) \equiv \int_0^\beta \langle A(-i\lambda)B(t)\rangle_0\ d\lambda, \tag{7.26}$$

and write (7.25) as

$$\chi_{BA}(\omega) = (A,\ B) - i\omega\int_0^\infty e^{i\omega t}(A,\ B(t))dt. \tag{7.27}$$

We see that

$$\chi_{BA}(0) = (A,\ B). \tag{7.28}$$

Equation (7.27) can then be written as

$$\chi_{BA}(\omega) = \chi_{BA}(0) - i\omega\int_0^\infty e^{i\omega t}R_{AB}(t)dt. \tag{7.29}$$

Using equation (7.23), we get

$$\langle[A,\ B(t)]\rangle = i\int_0^\beta d\lambda\langle A(-i\lambda)\dot{B}(t)\rangle_0, \tag{7.30}$$

which leads to

$$(A\dot{B}) = i\langle[A,\ B]\rangle. \tag{7.31}$$

7.2 Relation between susceptibility and Green function

The retarded one-particle Green function is defined by

$$G_{BA}(t - t') = -i\theta(t - t')\langle[B(t - t'),\ A]\rangle, \tag{7.32}$$

where the step function $\theta(t)$ is defined by

$$\theta(t) \equiv \begin{cases} 1 & \text{for } t > 0 \\ 0 & \text{for } t < 0 \end{cases}.$$

Note that some authors use i in place of $-i$ in the definition of the Green function.

We can write equation (7.15) as

$$\begin{aligned} \Delta B &= itr\int_{-\infty}^{t} \rho_0[B(t-t'), A]f(t')dt' \\ &= itr\int_{-\infty}^{\infty} \rho_0\theta(t-t')[B(t-t'), A]f(t')dt', \end{aligned} \tag{7.33}$$

or

$$\Delta B = -\int_{-\infty}^{\infty} G_{BA}(t-t')f(t')dt'. \tag{7.34}$$

If $f(t) = e^{-i\omega t}f$ we get

$$\chi_{BA}(\omega) = -\int_{-\infty}^{\infty} G_{BA}(t-t')e^{i\omega(t-t')}dt', \tag{7.35}$$

which leads to

$$\chi_{BA}(\omega) = -G_{BA}(\omega). \tag{7.36}$$

The inverse transform of $G_{BA}(\omega)$ is given by

$$G_{BA}(t) = \frac{1}{2\pi}\int_{-\infty}^{\infty} G_{BA}(\omega)e^{-i\omega t}d\omega. \tag{7.36b}$$

7.3 Correlation functions

Most of the experiments in physics are conveniently discussed in terms of correlation functions. In systems that are near thermal equilibrium, they all give desirable information about the intrinsic statistical fluctuations. When one probes dynamical processes, one usually employs an external field that disturbs the system slightly from equilibrium, and one measures the linear time-dependent response to this field. In all these experiments, what is probed is the dynamical behavior of spontaneous fluctuations about the equilibrium state. These fluctuations, according to the linear response theory, can be described in terms of time-dependent correlation functions, defined by

$$S_{AB}(t, t') \equiv \langle A(t)B(t')\rangle - \langle A(t)\rangle\langle B(t')\rangle, \tag{7.37}$$

where we have subtracted the equilibrium averages, so that $S_{AB}(t) = S_{AB}(t, 0) \to 0$ as $t \to \infty$, and its Fourier transform $S_{AB}(\omega)$ is well defined. Of course $\langle A(t)\rangle$ is time-independent.

Using $\rho = e^{-\beta H}/Z$ we find

$$tre^{-\beta H}A(t)B(0) = trA(t + i\beta)e^{-\beta H}B(0) = tre^{-\beta H}B(0)A(t + i\beta). \tag{7.38}$$

Time translation invariance implies $\langle A(t)B(0)\rangle = \langle A(0)B(-t)\rangle$, and therefore

$$\begin{aligned} S_{BA}(-t) &= \langle B(-t)A(0)\rangle = \langle B(0)A(t)\rangle \\ &= tre^{-\beta H}A(0)B(-t + i\beta) = \langle A(t - i\beta)B(0)\rangle, \end{aligned} \tag{7.39}$$

or

$$S_{BA}(-t) = S_{AB}(t - i\beta) = e^{-i\beta\partial_t}S_{AB}(t), \tag{7.40}$$

(where we have used the well-known result $f(x - a) = e^{-ad/dx}f(x)$ [1]).

We introduce now the Laplace transform $K(t)$ by (the usual s is replaced by $-iz$)

$$\tilde{\chi}(z) = \int_0^\infty K(t)e^{izt}dt, \quad \text{for Im } z > 0. \tag{7.41}$$

$\tilde{\chi}(z)$ is an analytic function in the upper part of the complex z-plane. $\chi(\omega)$ given by equation (7.21), is the limit of $\tilde{\chi}(z)$ as we approach the real frequency axis from above

$$\chi(\omega) = \lim_{\varepsilon\to 0^+}\tilde{\chi}(\omega + i\varepsilon) = \chi'(\omega) + i\chi''(\omega). \tag{7.42}$$

From this equation, we get

$$2i\chi''(\omega) = \chi(\omega + i\varepsilon) - \chi(\omega - i\varepsilon). \tag{7.43}$$

If the system is translationally invariant in time, we can define a Fourier transform by

$$K(t) = \frac{1}{2\pi}\int_{-\infty}^\infty e^{-i\omega t}K(\omega)d\omega. \tag{7.44}$$

Then

$$\tilde{\chi}(z) = \frac{1}{2\pi}\int_0^\infty dt\int_{-\infty}^\infty e^{-i(\omega-z)t}K(\omega)d\omega = \frac{1}{2\pi i}\int_{-\infty}^\infty \frac{K(\omega)d\omega}{\omega - z}. \tag{7.45}$$

Now

$$\tilde{\chi}(\omega + i\varepsilon)_{\varepsilon\to 0} = \frac{1}{2\pi i}\int_{-\infty}^\infty \frac{K(\omega')d\omega'}{\omega' - \omega - i\varepsilon}. \tag{7.46}$$

To calculate $\tilde{\chi}(\omega + i\varepsilon)$ it is better to write $K(t) = ix(t)$. Then using equation (7.82), we find

$$\chi''(\omega) = x(\omega) \text{ and } \chi'(\omega) = \frac{1}{\pi}P\int_{-\infty}^\infty \frac{\chi''(\omega')d\omega'}{\omega' - \omega}. \tag{7.47}$$

The connection between the real and imaginary parts of any complex function analytic in the upper half-plane is called the Kramers–Kronig relation. From equation (7.17) we obtain

$$\chi_{BA}(t) = \frac{1}{2}\langle[B(t), A]\rangle. \tag{7.48}$$

Using equations (7.48) and (7.40), we get

$$2\chi''_{AB}(t) = S_{AB}(t) - S_{BA}(-t) = (1 - e^{-i\beta\partial_t})S_{AB}(t). \tag{7.49}$$

Taking the Fourier transform of (7.49), using $\partial_t \to -i\omega$ (this is not a rigorous procedure, but the final expression is correct. We can also expand equation (7.40) in a Taylor series and calculate the Fourier transform of each term) we obtain

$$\chi''_{AB}(\omega) = \frac{1}{2}(1 - e^{-\beta\omega})S_{AB}(\omega). \tag{7.50}$$

The result (7.50) is known as the *Fluctuation-dissipation theorem*. It relates, for a system in thermal equilibrium, two distinct quantities: The spontaneous fluctuations described by $S_{AB}(\omega)$ which arise from the thermal motion of the particles of the system (even in the absence of external perturbation) and the dissipative behavior, on the other hand, described by $\chi''(\omega)$, which occur when the system is perturbed. Using equation (7.27), we can show that

$$\chi''_{AB}(\omega) = \frac{\omega}{2}R_{AB}(\omega), \tag{7.51}$$

where

$$R_{AB}(\omega) = \int_{-\infty}^{\infty} e^{i\omega t}(A, B(t))dt = 2\,\mathrm{Re}\int_{0}^{\infty} e^{i\omega t}(A, B(t))dt. \tag{7.52a}$$

Using equations (7.50) and (7.51), we obtain

$$R_{AB}(\omega) = \frac{(1 - e^{-\beta\omega})}{\omega}S_{AB}(\omega). \tag{7.52b}$$

Let us comment briefly on integral transforms. For instance, $S(t)$ being a fluctuation, there is nothing special about time larger than or smaller than zero. The behavior of $S(t)$ does not distinguish $t > 0$ from $t < 0$, and we can use Fourier transform or Laplace transform. However, if we have a perturbation applied to the system, for instance, a magnetic field starting $t = -\infty$ and suddenly turned off at $t = 0$, the behavior of the system is different from $t < 0$ and for $t > 0$. The magnetization, for instance, is constant for $t < 0$ but relaxes to zero after the field is turned off. In this case, we use the Laplace transform in time to represent the magnetization's relaxation behavior.

7.4 Sum rules

From equation (7.48) we obtain

$$i\frac{\partial^n}{\partial t^n}\chi''_{BA}(t) = \frac{1}{2}\left\langle\left[i\frac{\partial^n}{\partial t^n}B(t),\, A\right]\right\rangle. \tag{7.53}$$

Taken at time $t = 0$, that means that

$$\int_{-\infty}^{\infty}\frac{d\omega}{\pi}\omega^n\chi''_{BA}(\omega) = \langle[i^nB^{(n)},\, A]\rangle, \tag{7.54}$$

where

$$i^nB^{(n)} \equiv \left(i\frac{\partial}{\partial t}\right)^n B(t)|_{t=0} = [\dots[B(0),\, H]\dots,\, H]. \tag{7.55}$$

Thus, the right-hand side of (7.54) contains a sequence of equal time commutators, which can, in principle, be exactly calculated. In general, a sum rule is an exact result of the frequency integral of a certain frequency-dependent quantity. When we do calculations, usually we can only get some approximate values for these quantities, which may not satisfy precisely the sum rule. The difference between the two values can be taken as the quality of our approximation.

7.5 A simple example

To illustrate the formalism presented above, let us consider a Hamiltonian of the form

$$H = \sum_q \omega_q a_q^\dagger a_q. \tag{7.56}$$

The time dependence of the operators a_q and $a_q^\dagger$ are given by

$$a_q(t) = a_q e^{-i\omega_q t}, \qquad a_q^\dagger(t) = a_q^\dagger e^{i\omega_q t}. \tag{7.57}$$

Suppose we want to treat an operator defined by

$$A_q \equiv a_q^\dagger + a_q. \tag{7.58}$$

Note that $A_q^\dagger = A_q$ and therefore A_q is Hermitian. Using $\langle a_q a_q\rangle = \langle a_q^\dagger a_q^\dagger\rangle = 0$, the correlation function can be written as

$$S(t) = \langle A_q(t)A_q\rangle = \left\langle a_q^\dagger(t)a_q\right\rangle + \left\langle a_q(t)a_q^\dagger\right\rangle. \tag{7.59}$$

It follows that

$$S(t) = e^{i\omega_q t}\left\langle a_q^\dagger a_q\right\rangle + e^{-i\omega_q t}\left\langle a_q a_q^\dagger\right\rangle = n_q e^{i\omega_q t} + (1 + n_q)e^{-i\omega_q t}. \tag{7.60}$$

The Fourier transform is

$$S(\omega) = \int_{-\infty}^{\infty} e^{i\omega t} S(t)dt = n_q\delta(\omega + \omega_q) + (1 + n_q)\delta(\omega - \omega_q). \tag{7.61}$$

However, if we write $\omega_q \to \omega_q + i\varepsilon$, the above expression becomes

$$S(\omega) = \frac{n_q}{\omega + \omega_q + i\varepsilon} + \frac{1 + n_q}{\omega - \omega_q - i\varepsilon}. \tag{7.62}$$

Or, using $n_q = (e^{\beta\omega_q} - 1)^{-1}$ we obtain

$$S(\omega) = \frac{1}{e^{\beta\omega_q} - 1}\left[\frac{1}{\omega + \omega_q + i\varepsilon} + \frac{e^{\beta\omega_q}}{\omega - \omega_q - i\varepsilon}\right]. \tag{7.63}$$

The symmetry is broken between $\omega = \omega_q$ and $\omega = -\omega_q$ because at $T = 0$, the system is in the ground state. It can be excited, but it cannot decay. The relaxation function is given by

$$\begin{aligned}
R(t) &= \int_0^\beta \langle A_q(t - i\lambda)A_q\rangle \, d\lambda \\
&= \int_0^\beta [n_q e^{i\omega_q(t-i\lambda)} + (1 + n_q)e^{-i\omega_q(t-i\lambda)}]d\lambda \\
&= n_q e^{i\omega_q t}\int_0^\beta d\lambda e^{\omega_q\lambda} + (1 + n_q)e^{-i\omega_q t}\int_0^\beta d\lambda e^{-\omega_q\lambda} \\
&= \frac{e^{i\omega_q t}}{\omega_q} + \frac{e^{-i\omega_q t}}{\omega_q}(1 + n_q)(1 - e^{-\beta\omega_q}).
\end{aligned} \tag{7.64}$$

Using $1 + n_q = e^{\beta\omega_q}n_q$ we have $(1 + n_q)(1 - e^{-\beta\omega_q}) = 1$, and hence

$$R(t) = \frac{e^{i\omega_q t} + e^{-i\omega_q t}}{\omega_q}. \tag{7.65}$$

Note that $R(t)$ does not describe spontaneous fluctuations, but the relaxation after the system was perturbated, and so there is no Bose factor.

7.6 Spin transport

Up to now, for simplicity, I have not considered the spatial dependence of the operators. We can repeat all the previous sections' procedures, including a dependence with $\vec{r}$ and taking Fourier transform in the spatial variable. In this section, I will consider the particular case of a magnetic system with the spin S^z interacting with an external magnetic field $B(\vec{r})$ along the z-direction. It is less cumbersome to study the interaction term in the continuum limit, instead of a discrete lattice (the result will be independent of the method used) and write

$$V(t) = \int S^z(\vec{r}, t)B(\vec{r})d\vec{r}. \tag{7.66}$$

Differentiating with respect to t we find

$$\frac{dV(t)}{dt} = \int \frac{dS^z(\vec{r}, t)}{dt} B(\vec{r}) d\vec{r}. \tag{7.67}$$

If the S^z component commutates with the Hamiltonian, we have the continuity equation

$$\dot{S}^z(\vec{r}, t) + \vec{\nabla} \cdot \vec{j}(\vec{r}, t) = 0, \tag{7.68}$$

where $\vec{j}(\vec{r}, t)$ is the spin current. Taking (7.68) into (7.67) we get

$$\frac{dV(t)}{dt} = \int [-\vec{\nabla} \cdot \vec{j}(\vec{r}, t)] B(\vec{r}) d\vec{r} = -\int \vec{j}(\vec{r}, t). \vec{\nabla} B(\vec{r}) d\vec{r}, \tag{7.69}$$

where we have integrated by parts to obtain the last term. The response to the spin interaction with the magnetic field is the spin current, and we have from (7.13)

$$\langle \vec{j}(\vec{r}, t) \rangle = tr[\Delta\rho \vec{j}(\vec{r}, t)], \tag{7.70}$$

and equation (7.11) takes the form

$$\Delta\rho = \rho_0 \int_0^\infty dt \int_0^\beta d\lambda \int d\vec{r} \vec{j}(\vec{r}, -t - i\lambda) \cdot \vec{\nabla} B(\vec{r}). \tag{7.71}$$

Taking (7.71) into (7.70), and Fourier transforming leads to

$$\langle j_{q\alpha} \rangle = \int_0^\infty dt \int_0^\beta d\lambda \langle j_{-q\alpha}(-t - i\lambda) j_{q\gamma}(t) \rangle \nabla_{q\gamma} B, \tag{7.72}$$

where $\alpha, \gamma = x, y, z$. The conductivity tensor $\sigma_{\alpha\gamma}$ defined by $\langle j_{q\alpha} \rangle = \sigma_{\alpha\gamma}(\omega) \nabla_{q\gamma} B$ is obtained from (7.72) by letting $\vec{q}$ go to zero. If $j_{q\alpha}(t) = j_{q\alpha} e^{i\omega t}$ we find

$$\sigma_{\alpha\gamma}(\omega) = \lim_{s \to 0} \frac{1}{V} \int_0^\infty e^{-\eta t} dt \int_0^\beta d\lambda \langle j_\alpha(-t - i\lambda) j_\gamma \rangle e^{i\omega t}, \tag{7.73}$$

where here V is the volume of the system, j_γ is evaluated at $q = 0$, and we have reintroduced the factor $e^{-\eta t}$. The spin current is a response to an external magnetic field gradient. Equation (7.73) is sometimes written as

$$\sigma_{\alpha\beta}(\omega) = \frac{1}{V} \int_0^\beta d\lambda \int_0^\infty e^{i\omega t} \langle j_\alpha(-i\lambda) j_\beta(t) \rangle \, dt. \tag{7.74}$$

Using a positive sign for the perturbation term, we can rewrite equation (7.27) as

$$\int_0^\infty e^{i\omega t} R_{AB}(t) dt = \frac{i}{\omega}(A, B) + \frac{1}{\omega} \int_0^\infty e^{i\omega t} \langle [B(t), A] \rangle \, dt. \tag{7.75}$$

Using (7.75) we see that equation (7.74) has the equivalent form

$$\sigma_{\alpha\beta}(\omega) = \frac{i}{V\omega}(j_\alpha, j_\beta) + \frac{1}{V\omega} \int_0^\infty e^{i\omega t} \langle [j_\alpha(t), j_\beta(0)] \rangle \, dt, \tag{7.76}$$

or

$$\sigma_{\alpha\beta}(\omega) = \frac{i}{V\omega}(j_\alpha, j_\beta) + \frac{1}{\omega i}\chi_{\alpha\beta}(q = 0, \omega). \tag{7.77}$$

The first term in equation (7.76) is singular, diverging as $\omega \to 0$. This divergence cancels against pieces from the second term. The version (7.74) gets rid of the commutator and singular first term, at the cost of introducing an imaginary time $-i\lambda$.

7.7 Kubo formulas

The averages in equations (7.74) and (7.76) can be evaluated using the complete set of eigenstates $\{|n\rangle\}$ of H_0 and the associated eigenvalues E_n, that is $H_0|n\rangle = E_n|n\rangle$, where, as was said before, H_0 is the Hamiltonian before the perturbation is applied. Note that it may happen that we do not know what these eigenstates are, but we know that they exist. For our purpose, this is sufficient. Inserting a resolution of the identity $I = \sum_m |m\rangle\langle m|$ between the operators j_α and j_β, and performing the integrals, we can write equations (7.74) and (7.76) as

$$\sigma_{\alpha\beta}(\omega) = \frac{i}{V}\sum_{m,n}\frac{e^{-\beta E_n} - e^{-\beta E_m}}{Z(E_m - E_n)}\frac{\langle n|\, j_\alpha\,|m\rangle\langle m|\, j_\beta\,|n\rangle}{\omega + i\eta - (E_m - E_n)}, \tag{7.78}$$

$$\sigma_{\alpha\beta}(\omega) = \frac{i}{\omega}(j_\alpha, j_\beta) + \frac{i}{V\omega}\sum_{m,n}\frac{e^{-\beta E_n} - e^{-\beta E_m}}{Z}\frac{\langle n|j_\alpha|m\rangle\langle m|\, j_\beta\,|n\rangle}{\omega + i\eta - (E_m - E_n)}, \tag{7.79}$$

where I have added the positive infinitesimal η ($\omega \to \omega + i\eta$) to make the time integrals well defined. Note that $(j_\alpha, j_\beta) = 0$ if $\alpha \neq \beta$. Another way to obtain (7.79) from (7.78) is as follows. From equation (7.26), we have

$$(j_\alpha, j_\beta) = -\sum_{m,n}\frac{e^{-\beta E_m} - e^{-\beta E_n}}{Z}\frac{\langle m|j_\alpha|n\rangle\langle n|\, j_\beta\,|m\rangle}{E_m - E_n}. \tag{7.80}$$

Writing

$$\frac{1}{E_{mn} \pm \Omega} = \frac{1}{E_{mn}}\left(1 \pm \frac{\Omega}{E_{mn} + \Omega}\right), \tag{7.81}$$

where $\Omega = \omega + i\eta$, $E_{mn} = E_m - E_n$, and substituting these expressions into (7.79), the first term of the decomposition cancels the term (7.80), and we get (7.78). If $\langle m|j_\alpha|n\rangle\langle n|j_\beta|m\rangle$ is real, using

$$\frac{1}{\omega - E + i\eta} = P\left(\frac{1}{\omega - E}\right) - i\pi\delta(\omega - E), \tag{7.82}$$

we have for the real part of (7.78)

$$\mathrm{Re}\,\sigma_{\alpha\beta}(\omega) = -\frac{\pi}{V}\sum_{mn}\frac{e^{-\beta E_n} - e^{-\beta E_m}}{Z(E_n - E_m)}\langle n|j_\alpha|m\rangle\langle m|j_\beta|n\rangle\delta(\omega - (E_m - E_n)). \tag{7.83}$$

Equation (7.83) can be rewritten in terms of a time-dependent current–current correlation function as

$$\mathrm{Re}\,\sigma_{\alpha\beta}(\omega) = \frac{1 - e^{-\beta\omega}}{2\omega V}\int_{-\infty}^{\infty}\langle j_\alpha(t)j_\beta(0)\rangle\, dt. \tag{7.84}$$

Let $p_m = e^{-\beta E_m}$. For $E_m \to E_n$, it follows that

$$p_m = p_n + \frac{\partial p_n}{\partial E}(E_n - E_m) = p_n - \beta e^{-\beta E_n}(E_n - E_m). \tag{7.85}$$

Then we can rewrite (7.83) as

$$\begin{aligned}\mathrm{Re}\,\sigma_{\alpha\beta}(\omega) = &\frac{\beta\pi}{V}\sum_{n,m}\frac{e^{-\beta E_n}}{Z}\langle n|j_\alpha|m\rangle\langle m|j_\beta|n\rangle\delta(\omega)\\ &+\frac{\pi}{V}\sum_{mn}\frac{e^{-\beta E_n} - e^{-\beta E_m}}{Z(E_n - E_m)}\langle n|j_\alpha|m\rangle\langle m|j_\beta|\\ &n\rangle\delta(\omega - (E_m - E_n)).\end{aligned} \tag{7.86}$$

In the first term we have $E_n = E_m$, and in the second $E_n \neq E_m$. equation (7.86) is usually decomposed as

$$\mathrm{Re}\,\sigma_{\alpha\beta}(\omega) = D_{\alpha\beta}\delta(\omega) + \sigma_{\alpha\beta}^{reg}(\omega). \tag{7.87}$$

The coefficient of the zero frequency δ function contribution is called the Drude weight and is given by

$$D_{\alpha\beta} = \frac{\beta\pi}{V}\sum_{n,m}\frac{e^{-\beta E_n}}{Z}\langle n|j_\alpha|m\rangle\langle m|j_\beta|n\rangle. \tag{7.88}$$

In this equation $E_m = E_n$. This term is interpreted as the contribution of thermally excited particles that propagate ballistically without any interactions with other particles, i. e., a diverging d.c. conductivity. Scattering processes are expected to lead to a temperature-dependent broadening of the delta peak. The regular part of the conductivity can be written as

$$\sigma_{\alpha\beta}^{reg}(\omega) = \frac{\pi}{V\omega}\sum_{mn}\frac{e^{-\beta E_n} - e^{-\beta E_m}}{Z}\langle n|j_\alpha|m\rangle\langle m|j_\beta|n\rangle\delta(\omega - (E_m - E_n)). \tag{7.89}$$

We can demonstrate the following sum rule

$$\frac{2}{\pi}\int_0^{\infty} d\omega\,\mathrm{Re}\,\sigma_{\alpha\beta}(\omega) = (j_\alpha, j_\beta). \tag{7.90}$$

In the noninteracting approximation, where we have free particles (magnons in our case), the current operator can be written as [2]:

$$\hat{j}_x(t) = \sum_{n,m} \langle n | j_x | m \rangle c_n^\dagger c_m e^{i(\varepsilon_n - \varepsilon_m)t}, \tag{7.91a}$$

where $H_0|n\rangle = \varepsilon_n|n\rangle$ and c_n are boson operators. Using

$$\begin{aligned} tr\ \rho_0\left(\left[c_n^\dagger c_m,\ c_p^\dagger c_q\right]\right) = \delta_{mp}\delta_{nq}(n_n - n_m), \quad & tr\ \rho_0\left(c_n^\dagger c_m c_p^\dagger c_q\right) \\ & = \delta_{mp}\delta_{nq} n_n(1 - n_m) + \delta_{nm}\delta_{pq} n_n n_p \end{aligned} \tag{7.91b}$$

where $n_n = (e^{\beta\varepsilon_n} - 1)^{-1}$, we find from (7.78)

$$\sigma_{\alpha\beta}(\omega) = \frac{i}{V}\sum_{n,m}\left(\frac{n_n - n_m}{\varepsilon_m - \varepsilon_n}\right)\frac{\langle n | j_\alpha | m \rangle\langle m | j_\beta | n \rangle}{\omega + i\eta + \varepsilon_n - \varepsilon_m}. \tag{7.92}$$

If we take into account the spatial dependence of the system, equation (7.92) would become

$$\begin{aligned} \sigma_{\alpha\beta}(\vec{q},\ \omega) = & -\frac{i}{V}\sum_{n,m}\sum_{k}\left(\frac{n_{n,k} - n_{m,k+q}}{\varepsilon_{n,k} - \varepsilon_{m,k+q}}\right) \\ & \times \frac{\langle n,\ \vec{k} | j_{k,\alpha} | m,\ \vec{k} + \vec{q} \rangle\langle m,\ \vec{k} + \vec{q} | j_{k+q,\beta} | n,\ \vec{k} \rangle}{\omega + i\eta + \varepsilon_{n,k} - \varepsilon_{m,k+q}} \end{aligned} \tag{7.93}$$

and similarly for other expressions.

As a simple example of using the procedure described above, I will calculate $\sigma_{xx}^{reg}(\vec{q} = 0,\ \omega)$ for the noninteracting antiferromagnet described by the Hamiltonian (2.31). The energy is given by $E_\pm(\vec{k}) = \pm\omega_k$, with $\omega_k = JSz\lambda_k$, and $\lambda_k = \sqrt{1 - \gamma_k^2}$. Now we have to consider both signs of the energy. From (4.37) we get the eigenvectors

$$|+,\ \vec{k}\rangle = \begin{pmatrix} u_k \\ v_k \end{pmatrix}, \qquad |-,\ \vec{k}\rangle = \begin{pmatrix} v_k \\ u_k \end{pmatrix}. \tag{7.94}$$

Using (2.34) and (2.39) we can write

$$u_k = \sqrt{\frac{1 + \lambda_k}{2\lambda_k}}, \qquad v_k = \sqrt{\frac{1 - \lambda_k}{2\lambda_k}}. \tag{7.95}$$

In analogy with charge transport, we define the spin current for noninteracting magnons as

$$j_{k,x} = \frac{\partial M_k}{\partial k_x}, \tag{7.96}$$

where M_k was defined in equation (4.19). From (2.31) we can write an expression for the matrix M_k and get

$$j_{k,x} = \frac{JSz}{2}\begin{pmatrix} 0 & \sin k_x \\ \sin k_x & 0 \end{pmatrix}. \tag{7.97}$$

Using (7.97) and (7.94) we obtain

$$\langle +, \vec{k}\,|j_{k,x}|-, \vec{k}\rangle = \frac{JSz}{2}(u_k^2 + v_k^2)\sin k_x = \frac{JSz}{2\lambda_k}\sin k_x. \tag{7.98}$$

We can then write using equation (7.93)

$$\sigma_{xx}^{reg}(\vec{q} = 0, \omega) = \pi\frac{(JSz)^2}{4V}\sum_k \frac{\sin^2 k_x[n(E_+) - n(E_-)]}{\lambda_k^2[E_+(\vec{k}) - E_-(\vec{k})]}\delta(\omega - (E_+ - E_-)). \tag{7.99}$$

Using the relation $n(-x) = -1 - n(x)$, we get the final expression

$$\sigma_{xx}^{reg}(\vec{q} = 0, \omega) = \pi\frac{(JSz)^4}{8V}\sum_k \frac{\sin^2 k_x}{\omega_k^3}(1 + 2n(\omega_k))\delta(\omega - 2\omega_k). \tag{7.100}$$

Equation (7.100) agrees with the result obtained by Sentef [3] in the limit of noninteracting magnons using Green function theory. In chapter 9, I will present the calculation of σ_{xy} for a more complex model.

7.8 Green functions

We define for a general bosonic operator $O(\vec{r}, t)$ the following functions:

$$G_>(\vec{r} - \vec{r}\,', t) = \langle O(\vec{r}, t)O(\vec{r}\,', 0)\rangle, \tag{7.101}$$

$$G_<(\vec{r} - \vec{r}\,', t) = \langle O(\vec{r}\,', 0)O(\vec{r}, t)\rangle, \tag{7.102}$$

$$G^R(\vec{r} - \vec{r}\,', t) = -i\theta(t)\langle[O(\vec{r}, t), O(\vec{r}\,', 0)]\rangle, \tag{7.103}$$

$$G_F(\vec{r} - \vec{r}\,', t) = -i\langle T(O(\vec{r}, t))(\vec{r}\,', 0))\rangle, \tag{7.104}$$

where Wick's time ordering operator T is defined by

$$T(A(t)B(t')) = \theta(t - t')A(t)B(t') + \theta(t' - t)B(t')A(t). \tag{7.105}$$

For $G_>$ and $G_<$, we have only a thermal average of a product (not necessarily ordered) of Heisenberg operators. G^R is the retarded Green function, introduced in section 7.2. G_F is the time-ordered Green function, also known as the Feynman propagator. This function will be studied in chapter 8 using perturbation theory.

Let us first consider the first function. We can write

$$\begin{aligned}\langle O(\vec{r}, t)O(\vec{r}\,', 0)\rangle &= \frac{tr(e^{-\beta H}e^{iHt}O(\vec{r})e^{-iHt}O(\vec{r}\,'))}{tre^{-\beta H}} \\ &= \frac{tr(O(\vec{r}\,')e^{-\beta H}e^{iHt}O(\vec{r})e^{-iHt})}{tre^{-\beta H}} \\ &= \frac{1}{Z}\sum_{m,n}\langle n|O(\vec{r}\,')e^{-\beta H}|m\rangle\langle m|e^{iHt}O(\vec{r})e^{-iHt}|n\rangle \\ &= \frac{1}{Z}\sum_{nm}e^{-\beta E_m}\langle n|O(\vec{r}\,')|m\rangle\langle m|O(\vec{r})|n\rangle e^{i(E_m - E_n)t}\end{aligned} \tag{7.106}$$

where I have used the property $tr(AB) = tr(BA)$. The time Fourier transform of (7.101) is

$$\begin{aligned} J_1(\vec{r} - \vec{r}', \omega) &= \int_{-\infty}^{\infty} \langle O(\vec{r}', t)O(\vec{r}, 0)\rangle e^{i\omega t}dt \\ &= \frac{2\pi}{Z}\sum_{nm} e^{-\beta E_m}\langle n|O(\vec{r}')|m\rangle\langle m|O(\vec{r})|n\rangle\delta(E_m - E_n + \omega). \end{aligned} \tag{7.107}$$

J_1 is called the spectral function of the time correlator $\langle O(\vec{r}, t)O(\vec{r}', 0)\rangle$.

Following the same procedure, we arrive at the time Fourier transform of $G_<$ and get

$$\begin{aligned} J_2(\vec{r} - \vec{r}', \omega) &= \int_{-\infty}^{\infty} \langle O(\vec{r}', t)O(\vec{r}, 0)\rangle e^{i\omega t}dt \\ &= \frac{2\pi}{Z}\sum_{nm} e^{-\beta En}\langle n|O(\vec{r}')|m\rangle\langle m|O(\vec{r})|n\rangle\delta(E_m - E_n + \omega). \end{aligned} \tag{7.108}$$

The Dirac delta function gives $E_m = E_n - \omega$, hence

$$\begin{aligned} J_2(\vec{r} - \vec{r}', \omega) &= \frac{2\pi}{Z}\sum_{nm} e^{-\beta(E_m+\omega)}\langle n|O(\vec{r}')|m\rangle\langle m|O(\vec{r})|n\rangle\delta(E_m - E_n + \omega) \\ &= e^{-\beta\omega}J_1(\vec{r} - \vec{r}', \omega). \end{aligned} \tag{7.109}$$

Now consider the retarded Green function

$$\begin{aligned} G^R(\vec{r} - \vec{r}', t) &= -i\theta(t)(\langle O(\vec{r}, t)O(\vec{r}', 0) - O(\vec{r}', 0)O(\vec{r}, t)\rangle) \\ &= -i\theta(t)(G_>(\vec{r} - \vec{r}', t) - G_<(\vec{r} - \vec{r}', t)). \end{aligned} \tag{7.110}$$

For a translationally invariant system, we can express the retarded Green function in terms of the Fourier transformed spectral functions $J_1(\vec{k}, \omega)$ and $J_2(\vec{k}, \omega)$. Using equations (7.107) and (7.109) we obtain

$$G^R(\vec{k}, \omega) = \int_{-\infty}^{\infty} \frac{d\omega'}{2\pi}(1 - e^{-\beta\omega'})\frac{J_1(\vec{k}, \omega')}{\omega - \omega' + i\eta}. \tag{7.111}$$

This function has poles in the lower half-plane Im $\omega < 0$. Using the relation

$$\lim_{\eta\to 0^+} \frac{1}{x + i\eta} = \wp\left(\frac{1}{x}\right) - i\pi\delta(x), \tag{7.112}$$

we get

$$\begin{aligned} \text{Im } G^R(\vec{k}, \omega) &= -\pi\int_{-\infty}^{\infty} \frac{d\omega'}{2\pi}(1 - e^{-\beta\omega'})J_1(\vec{k}, \omega')\delta(\omega - \omega') \\ &= -\frac{1}{2}(1 - e^{-\beta\omega})J_1(\vec{k}, \omega), \end{aligned} \tag{7.113}$$

which leads to

$$J_1(\vec{k}, \omega) = \left(\frac{2}{e^{-\beta\omega} - 1}\right)\text{Im } G^R(\vec{k}, \omega). \tag{7.114}$$

The Feynman propagator can be written as

$$\begin{aligned} G_F(\vec{r} - \vec{r}', t) = &- i\theta(t)\int_{-\infty}^{\infty} \frac{d\omega'}{2\pi} J_1(\vec{r} - \vec{r}', \omega')e^{-i\omega' t} \\ &+ i\theta(-t)\int_{-\infty}^{\infty} \frac{d\omega'}{2\pi} J_2(\vec{r} - \vec{r}', \omega')e^{-i\omega' t}. \end{aligned} \tag{7.115}$$

Taking the time Fourier transform, we find

$$\begin{aligned} G_F(\vec{k}, \omega) &= \int_{-\infty}^{\infty} \frac{d\omega'}{2\pi}\left\{\frac{J_1(\vec{k}, \omega')}{\omega - \omega' + i\eta} + \frac{J_2(\vec{k}, \omega')}{\omega - \omega' - i\eta}\right\} \\ &= \int_{-\infty}^{\infty} \frac{d\omega'}{2\pi} J_1(\vec{k}, \omega')\left\{\frac{1}{\omega - \omega' + i\eta} + \frac{e^{-\beta\omega'}}{\omega - \omega' - i\eta}\right\}. \end{aligned} \tag{7.116}$$

The expression (7.116) has poles both in the upper and lower half of the complex ω-plane. $J_1(\vec{k}, \omega)$ is real, which allows us to write

$$\text{Re } G^R(\vec{k}, \omega) = \text{Re } G_F(\vec{k}, \omega) = \wp\int_{-\infty}^{\infty} (1 - e^{-\beta\omega'})\frac{J_1(\vec{k}, \omega')}{\omega - \omega'}\frac{d\omega'}{2\pi}. \tag{7.117}$$

Following a similar procedure, we obtain

$$\text{Im } G^R(\vec{k}, \omega) = \tanh\left(\frac{\beta\omega}{2}\right)\text{Im } G_F(\vec{k}, \omega). \tag{7.118}$$

Note that at $T = 0$, $\text{Im } G^R(\vec{k}, \omega) = \text{Im } G_F(\vec{k}, \omega)$.

7.9 Equation of motion for the retarded Green function

It is usual in the literature to use the following notation when studying the equation of motion for the retarded Green function,

$$\langle\langle A(t); B(t')\rangle\rangle = -i\theta(t - t')\langle[A(t), B(t')]\rangle. \tag{7.119}$$

If $t' = 0$, the derivative of the above equation with respect to t is

$$\frac{d}{dt}\langle\langle A(t); B\rangle\rangle = -i\delta(t)\langle[A, B]\rangle - i\theta(t)\langle[[A(t), H], B]\rangle. \tag{7.120}$$

A Fourier transformation leads to the equation of motion

$$\omega\langle\langle A; B\rangle\rangle_\omega = \frac{1}{2\pi}\langle[A, B]\rangle + \langle\langle[A, H]; B\rangle\rangle_\omega. \tag{7.121}$$

The suffix ω indicates the Fourier transform. The last term in equation (7.121) is a new Green function (GF), and generally, this new term is of a higher order. A new GF is generated, of even higher order, and we get an infinite sequence. The appearance of an infinite chain of coupled equations of the (7.121) type is unavoidable for a system of interacting particles because one cannot consider a single particle or a group of particles independently of the rest of the system. We have to decouple a higher-order GF in terms of lower order to cut the series.

A and B can be the same Hermitian operator, in which case $[A, B] = 0$, and we proceed to the second derivative and obtain

$$\omega^2\langle\langle A; A\rangle\rangle_\omega = \langle [A, H], A\rangle + \langle\langle [A, H]; [A, H]\rangle\rangle, \tag{7.122}$$

where we have used

$$\langle [[[A, H], H], H]\rangle = -\langle [[A, H], [A, H]]\rangle. \tag{7.123}$$

The two-time Green function method is well known as the standard one, which gives consistent results over a wide temperature range. In applying this method, the so-called random phase approximation (RPA) or Tyablikov approximation is usually introduced to obtain the approximate solution from the infinite set of coupled equations in the GFs. The decoupling at a stage one-step further than the Tyablikov decoupling should be adopted when there is no long-range order [4].

Using equations (7.35), (7.43) and (7.50), we find that the static correlation function is given by

$$\langle AB\rangle = i\int_{-\infty}^{\infty} d\omega[\langle\langle A; B\rangle\rangle_{\omega+i\varepsilon} - \langle\langle A; B\rangle\rangle_{\omega-i\varepsilon}]\frac{1}{e^{\beta\omega} - 1}. \tag{7.124}$$

For instance, if

$$G(\omega) = \frac{\gamma}{\omega^2 - \omega_0^2} = \frac{\gamma}{(\omega + \omega_0)(\omega - \omega_0)}, \tag{7.125}$$

we have

$$\begin{aligned}\langle AB\rangle = i\gamma\int_{-\infty}^{\infty} d\omega \\ \times\Bigg[\frac{1}{(\omega + i\varepsilon + \omega_0)(\omega + i\varepsilon - \omega_0)} \\ - \frac{1}{(\omega - i\varepsilon + \omega_0)(\omega - i\varepsilon - \omega_0)}\Bigg]n(\omega).\end{aligned} \tag{7.126}$$

Integrating using residues, we find

$$\langle AB\rangle = i\gamma\left\{2\pi i\left[\frac{n(\omega_0)}{2\omega_0} - \frac{n(-\omega_0)}{2\omega_0}\right]\right\} = -\frac{2\pi\gamma}{2\omega_0}[n(\omega_0) - n(-\omega_0)], \tag{7.127}$$

or

$$\langle AB \rangle = \frac{\pi \gamma}{\omega_0} \coth\left(\frac{\beta \omega_0}{2}\right). \tag{7.128}$$

As an example, we will consider the three-dimensional Heisenberg ferromagnet

$$H = -\sum_{i \neq j} J_{ij} \vec{S}_i . \ \vec{S}_j, \tag{7.129}$$

below the critical temperature, as we need $\langle S^z \rangle \neq 0$ in the following procedure.

We want to calculate the Green function

$$G_{ij}(t) = \left\langle \left\langle S_i^+(t); \ S_j^- \right\rangle \right\rangle. \tag{7.130}$$

The Fourier transform is

$$G_{ij}(\omega) = \int_{-\infty}^{\infty} dt e^{i\omega t} G_{ij}(t). \tag{7.131}$$

The equation of motion becomes

$$\omega G_{ij}(\omega) = \left\langle \left[S_i^+, \ S_j^- \right] \right\rangle + \left\langle \left\langle [S_i^+, \ H]; \ S_j^- \right\rangle \right\rangle. \tag{7.132}$$

Using

$$\left[S_i^+, \ S_j^- \right] = 2\delta_{ij} S_i^z, \qquad [S_i^+, \ H] = -2\sum_m J_{mi}(S_m^+ S_i^z - S_i^+ S_m^z), \tag{7.133}$$

equation (7.132) can be written as

$$\omega G_{ij}(\omega) = 2\delta_{ij} \ \langle S_i^z \rangle - 2\sum_m J_{im} \left(\left\langle \left\langle S_i^z S_m^+; \ S_j^- \right\rangle \right\rangle_\omega - \left\langle \left\langle S_m^z S_i^+; \ S_j^- \right\rangle \right\rangle_\omega \right). \tag{7.134}$$

This equation is exact. However, it cannot be exactly solved since it contains higher-order Green functions. Therefore, we use the RPA decoupling

$$\begin{aligned} \left\langle \left\langle S_i^z S_m^+; \ S_j^- \right\rangle \right\rangle &\to \langle S_i^z \rangle \left\langle \left\langle S_m^+; \ S_j^- \right\rangle \right\rangle \\ \left\langle \left\langle S_m^z S_i^+; \ S_j^- \right\rangle \right\rangle &\to \langle S_m^z \rangle \left\langle \left\langle S_i^+; \ S_j^- \right\rangle \right\rangle. \end{aligned} \tag{7.135}$$

Since the system is translationally invariant, we find

$$\langle S_i^z \rangle = \langle S_m^z \rangle = \langle S^z \rangle. \tag{7.136}$$

We get for the equation of motion

$$(\omega - 2\langle S^z \rangle J_0) G_{ij}(\omega) = 2\delta_{ij} \langle S^z \rangle \sum_m J_{mj} G_{mj}(\omega). \tag{7.137}$$

Using the Fourier series

$$G_k(\omega) = \frac{1}{N}\sum_{ij} G_{ij}(\omega)e^{i\vec{k}.(\vec{r}_i-\vec{r}_j)}, \tag{7.138}$$

we get

$$G_k(\omega) = \frac{2\langle S^z\rangle}{\omega - \omega_k + i0^+}, \tag{7.139}$$

where

$$\omega_k = 2\langle S^z\rangle(J_0 - J(\vec{k})), \tag{7.140}$$

and

$$J(\vec{k}) = \frac{1}{N}\sum_{ij} J_{ij}e^{-i\vec{k}.(\vec{r}_i-\vec{r}_j)} = \sum_i J_{ij}e^{-i\vec{k}.(\vec{r}_i-\vec{r}_j)}. \tag{7.141}$$

Now I will show how to calculate $\langle S^z\rangle$. Using the following identity $S_i^y S_i^x + S_i^x S_i^y = 0$, which holds for $S = ½$, and the commutation relation $[S_i^x, S_i^y] = iS_i^z$ we get

$$\langle S_i^z\rangle = \frac{1}{2} - \langle S_i^- S_i^+\rangle. \tag{7.142}$$

From equations (7.142) and (7.124) we find

$$\langle S^z\rangle = \frac{1}{2} - \lim_{\varepsilon\to 0}\frac{1}{N}\sum_k \frac{i}{2\pi}\int_{-\infty}^{\infty} d\omega \frac{G_k(\omega + i\varepsilon) - G_k(\omega - i\varepsilon)}{e^{\beta\omega} - 1}. \tag{7.143}$$

Integrating using residues, we get

$$\langle S^z\rangle = \frac{1}{2} - \frac{1}{N}\sum_k \frac{2\langle S^z\rangle}{e^{\beta\omega_k} - 1} = \frac{1}{2} - \frac{1}{(2\pi)^3}\int d\vec{k}\frac{2\langle S^z\rangle}{e^{\beta\omega_k} - 1}, \tag{7.144}$$

where an integration over the Brillouin zone has replaced the sum over momenta. Using $2(e^x - 1)^{-1} = \coth(x/2) - 1$, we obtain the following expression for $\langle S^z\rangle$

$$\langle S^z\rangle = \left[\frac{2}{(2\pi)^2}\int d\vec{k}\,\coth\left(\frac{\beta\omega_k}{2}\right)\right]^{-1}. \tag{7.145}$$

Noting that $\langle S^z\rangle$ appear in ω_k, the above equation should be solved self consistently.

As another example, I will consider the Ising model in a transverse field in three dimensions. The Hamiltonian is

$$H = -\frac{1}{2}\sum_{ij} J_{ij}S_i^z S_j^z - \Omega\sum_i S_i^x. \tag{7.146}$$

From equation (7.120) we get the equations of motion

$$i\frac{d}{dt}\left\langle\left\langle S_i^z;\, S_j^z\right\rangle\right\rangle = -i\Omega\left\langle\left\langle S_j^y;\, S_i^z\right\rangle\right\rangle, \tag{7.147}$$

$$i\frac{d}{dt}\left\langle\left\langle S_i^y;\, S_j^z\right\rangle\right\rangle = i\langle S^x\rangle\delta_{ij}\delta(t) - i\sum_n J_{ni}\left\langle\left\langle S_n^z S_i^x;\, S_j^z\right\rangle\right\rangle + i\Omega\left\langle\left\langle S_i^z;\, S_j^z\right\rangle\right\rangle, \tag{7.148}$$

$$i\frac{d}{dt}\left\langle\left\langle S_i^x;\, S_j^z\right\rangle\right\rangle = i\sum_n J_{ni}\left\langle\left\langle S_n^z S_i^y;\, S_j^z\right\rangle\right\rangle. \tag{7.149}$$

In the next step, we perform the MFA decoupling

$$\left\langle\left\langle S_n^z S_i^x;\, S_j^z\right\rangle\right\rangle = \langle S^z\rangle\left\langle\left\langle S_i^x;\, S_j^z\right\rangle\right\rangle + \langle S^x\rangle\left\langle\left\langle S_n^z;\, S_j^z\right\rangle\right\rangle, \tag{7.150}$$

$$\left\langle\left\langle S_n^z S_i^y;\, S_j^z\right\rangle\right\rangle = \langle S^z\rangle\left\langle\left\langle S_i^y;\, S_j^z\right\rangle\right\rangle. \tag{7.151}$$

Taking the time Fourier transform, we find

$$\omega G_{ij}^{zz}(\omega) = -i\Omega G_{ij}^{yz}, \tag{7.152}$$

$$\omega G_{ij}^{yz}(\omega) = \frac{i}{2\pi}\langle S^x\rangle\delta_{ij} + i\Omega G_{ij}^{zz}(\omega) - i\langle S^z\rangle\sum_n J_{ni}G_{ij}^{yz} - i\langle S^x\rangle\sum_n J_{ni}G_{nj}^{zz}(\omega). \tag{7.153}$$

$$\omega G_{ij}^{xz}(\omega) = i\langle S^z\rangle\sum_n J_{ni}G_{ij}^{yz}(\omega). \tag{7.154}$$

Now we perform the space Fourier transform and obtain

$$\omega G_q^{zz}(\omega) = -i\Omega G_q^{yz}(\omega), \tag{7.155}$$

$$\omega G_q^{yz}(\omega) = \frac{i}{2\pi}\langle S^x\rangle + i\Omega G_q^{zz}(\omega) - iJ(0)\langle S^z\rangle G_q^{yz}(\omega) - iJ(\vec{q})\langle S^x\rangle G_q^{zz}(\omega), \tag{7.156}$$

$$\omega G_q^{xz}(\omega) = iJ(0)\langle S^z\rangle G_q^{yz}(\omega). \tag{7.157}$$

We get for $G_q^{zz}(\omega)$

$$G_q^{zz}(\omega) = \frac{1}{2\pi}\frac{\Omega\langle S^x\rangle}{\left(\omega^2 - \omega_q^2\right)}, \tag{7.158}$$

where

$$\omega_q^2 = \Omega^2 - \Omega J(\vec{q})\langle S^x\rangle + J^2(0)\langle S^z\rangle^2. \tag{7.159}$$

7.10 Green function in another context

The Green function concept is widely used in physics. I have presented some uses of it, and here I will show one more. We introduce the energy-dependent Green function through the equation

$$(E - H)G(E) = 1, \tag{7.160}$$

with the algebraic solution

$$G(E) = \frac{1}{E - H}. \tag{7.161}$$

We see that $G(E)$, considered as a function of E, has poles at the bound states of H.

The Fourier transform of (7.161) with $\omega = E$ is

$$G(t) = -ie^{-iHt}. \tag{7.162}$$

Suppose that we have the expression

$$S(k, \omega) = \int_{-\infty}^{\infty} dte^{i\omega t}\langle A_k(t)A_{-k}\rangle. \tag{7.163}$$

Using (7.162) we can write

$$S(k, \omega) = \int_{-\infty}^{\infty} dte^{i\omega t}\langle G(-t)A_k G(t)A_{-k}\rangle, \tag{7.164}$$

which can also be written as

$$S(k, \omega) = \int \frac{d\omega'}{2\pi}\langle G(\omega')A_k G(\omega' + \omega)A_{-k}\rangle. \tag{7.165}$$

If we have a Hamiltonian of the form

$$M(\vec{k}) = h_0(\vec{k}) + \vec{\sigma}\cdot\vec{h}(\vec{k}), \tag{7.166}$$

we can write $G(\vec{k}, \omega) = [\omega I - M(\vec{k})]^{-1}$. Calculating the inverse matrix, we obtain

$$G(\vec{k}, \omega) = \frac{1}{2h}\left[\frac{1 + \vec{\sigma}\cdot\vec{h}}{\omega - E_+(\vec{k})} - \frac{1 - \vec{\sigma}\cdot\vec{h}}{\omega - E_-(\vec{k})}\right], \tag{7.167}$$

where

$$E_\pm(\vec{k}) = h_0(\vec{k}) \pm h(\vec{k}), \qquad h = \sqrt{h_x^2 + h_y^2 + h_z^2}. \tag{7.168}$$

If $H = H_0 + V$, where V is a 'small' perturbation, we can write

$$G(E) = \frac{1}{E - H_0 - V} = \frac{1}{E - H_0} + \frac{1}{E - H_0}V\frac{1}{E - H_0} + \cdots \tag{7.169}$$

or

$$G(E) = G_0 + G_0VG_0 + \ldots = G_0 + G_0VG. \tag{7.170}$$

This equation is the so-called *Dyson equation* [5].

7.11 The memory function method

The memory function formalism is one of the best choice tools for describing non-quasiparticle physics in which a small number of slowly decaying variables dominate the dynamics. Our treatment here will follow Forster [6]. See also Florencio and Bonfin [7].

If $A(t)$ denotes a Hermitian dynamical variable, we can write the equation of motion

$$\frac{dA(t)}{dt} = iLA(t), \tag{7.171}$$

where L is the Liouville operator defined by $L = i[H, .]$, and H is the Hamiltonian of the system. Equation (7.171) has a formal solution

$$A(t) = e^{iLt}A(0). \tag{7.172}$$

We introduce the scalar product of two variables A and B by

$$(A(t), B) \equiv \int_0^\beta \langle e^{\lambda H}A(t)e^{-\lambda H}B\rangle \, d\lambda. \tag{7.173}$$

As we can see, the scalar product is the relaxation function introduced in section 7.1. Using (7.172) we can write

$$R(t) = (A(t), A(0)) = (A(0), A(-t)) = (A(0), e^{-iLt}A(0)). \tag{7.174}$$

The Laplace transform of equation (7.174) is given by the formal expression

$$R(z) = (A|\frac{i}{z - L}|A). \tag{7.175}$$

Now we define an operator that projects onto the vector $|A)$ as

$$P \equiv |A)(A|A^*)^{-1}(A| \equiv 1 - Q. \tag{7.176}$$

Note that although A is an operator, $|A)$ is considered as a vector in an enlarged vector space. We can, then, separate $A(t)$ into the projected and vertical components with respect to the A-axis (where $A = A(0)$).

$$A(t) = F(t)A + A'(t), \tag{7.177}$$

where

$$F(t) = R(t)/R(0), \tag{7.178}$$

and

$$A'(t) \equiv (1 - P)A(t) = QA(t). \tag{7.179}$$

Operating with Q on (7.171), and using (7.178) we get

$$\frac{dA'(t)}{dt} = iQLA'(t) + iF(t)QLA. \tag{7.180}$$

Let $L_1 = QL$. We integrate now (7.180), obtaining

$$A'(t) = \int_0^t F(s)e^{i(t-s)QL}L_1 A ds. \tag{7.181}$$

Let $g(t) = e^{itQL}L_1 A$, then

$$A'(t) = \int_0^t F(s)g(t - s)ds. \tag{7.182}$$

Let us split $\dot{A}$ into the projected and vertical components

$$\dot{A}(t) \equiv iLA = i\Omega A(t) + K(t), \tag{7.183}$$

where

$$i\Omega \equiv \left[\frac{dF(t)}{dt}\right]_{t=0} = (\dot{A}|A^*)(A, A^*)^{-1}, \qquad K = (1 - P)\dot{A} = Q\dot{A}. \tag{7.184}$$

From (7.178) and (7.183), we have

$$\frac{dF(t)}{dt} = i\Omega F(t) + (K(t), A^*)(A, A^*)^{-1}, \tag{7.185}$$

however,

$$(K(t), A^*) = (K, A^*(-t)) = (K, A'^*(-t)), \tag{7.186}$$

since K is orthogonal to A. Taking equation (7.182) into equation (7.186), we find

$$(K(t), A^*) = \int_0^{-t} (K, [F(s)g(-t - s)]^*)ds = \int_0^{-t} (K, g^*(-t - s))F^*(s)ds. \tag{7.187}$$

Using

$$F^*(t)(A, A^*)^{-1} = (A, A^*)^{-1}F(-t), \tag{7.188}$$

and $g(0) = K$, we can write (7.185) as

$$\frac{dF(t)}{dt} = i\Omega F(t) - \int_0^t \phi(t - \tau)F(\tau)d\tau, \qquad (t > 0), \tag{7.189}$$

where

$$\phi(t) \equiv (g, g(-t)^*)(A, A^*)^{-1}. \tag{7.190}$$

The function $\phi(t)$ is called the 'memory function': it describes the extent to which past values of $F(t)$ influence its time rate of change at present.

Using $g(t) = e^{itQL}QLA$, and $\dot{A} = LA$, we can write (7.190) as

$$\phi(t)(A, A^{*}) \equiv \gamma(t) = (\dot{A}|Qe^{-iQLQt}Q|\dot{A}), \tag{7.191}$$

where we have used the property $Q^2 = Q$. The Laplace transform of (7.191) is

$$\Sigma(z) = (A|LQ\frac{i}{z - LQ}L|A)(A, A^{*})^{-1}, \tag{7.192}$$

where $\Sigma(z)$ is the Laplace transform of the function $\phi(t)$. Finally, we take the Laplace transform of (7.189) and get

$$F(z) = \frac{i(A, A^{*})}{z - \Omega + i\Sigma(z)}. \tag{7.193}$$

If A has a definite time-reversal signature, the frequency Ω vanishes. We should be aware that this formalism is useful only if the memory function $\Sigma(z)$ is expected to be a simpler object than $F(z)$ itself. For some approximations of the memory function, see Pires [8] and Lee [9].

Mori [10] developed the memory function formalism in a slightly different approach, as shown in the following. He defined a hierarchy of generalized random forces $f_j(t)$ that obey the recurrence equation

$$f_j = \left(1 - \sum_{l=0}^{j-1} P_l\right) iLf_{j-1}, \qquad f_0 = A(0), \tag{7.194}$$

where P_l, the projection operator onto the vector f_l, is given by

$$P_l B \equiv (B, f_l)(f_l, f_l)^{-1} f_l, \tag{7.195}$$

and we are considering Hermitian operators here. Note that $f_1 = g$. These vectors f_j form an orthogonal set, $(f_j, f_k) = 0$ for $j \neq k$, and evolve in time according to the equation

$$f_j(t) = \exp(iL_j t) iL_j f_{j-1}, \quad j \geqslant 1, \quad f_0(t) = A(t), \tag{7.196}$$

where

$$L_j = (1 - P_{j-1})L_{j-1}, \qquad L_0 = L. \tag{7.197}$$

The initial time derivatives $\dot{f}_j = [df_j(t)/dt]_{t=0}$ satisfy the recurrence equations

$$\dot{f}_j = \left(1 - \sum_{l=0}^{j-1} P_l\right) iLf_j. \tag{7.198}$$

For Hermitian operators, we have

$$\dot{f}_j = f_{j+1}. \tag{7.199}$$

Introducing the quantities

$$\delta_j = (f_j, f_j)/(f_{j-1}, f_{j-1}), \tag{7.200}$$

the Laplace transform of the relaxation function $F(t)$ can be written

$$F(z) = \int_0^\infty F(t)e^{-zt}dt = \cfrac{1}{z + \cfrac{\delta_1}{z + \cfrac{\delta_2}{z + \dots}}} \tag{7.201}$$

This expansion is Mori's infinite continued fraction representation for the Laplace transform of the time relaxation function. Equation (7.201) can also be rewritten in the form

$$F(z) = \frac{1}{z + \delta_1 F_1(z)}, \qquad F_l(z) = \frac{1}{z + \delta_{l+1}F_{l+1}(z)}. \tag{7.202}$$

If $F(\omega)$ is the Fourier transform of $F(t)$, we define the moments $\langle\omega^{2n}\rangle$ by

$$\langle\omega^{2n}\rangle = \int_{-\infty}^{\infty} \omega^{2n}F(\omega)d\omega. \tag{7.203a}$$

It can be shown that

$$\begin{aligned} &\delta_1 = \langle\omega^2\rangle, \;\; \delta_2 = \langle\omega^4\rangle/\langle\omega^2\rangle - \langle\omega^2\rangle, \\ &\delta_3 = (\langle\omega^6\rangle/\langle\omega^2\rangle - \langle\omega^4\rangle^2/\langle\omega_2\rangle^2)\delta_2, \;\; \text{etc.} \end{aligned} \tag{7.203b}$$

If we go up to an appropriate $F_l(z)$, it is possible to introduce either one of the following approximations:

(a) A long-time approximation.
(b) A short-time approximation.
(c) A high- or low-temperature approximation.
(d) A perturbation calculation.

Now, I will present an approximation that we can use at low temperature when the Hamiltonian has the form

$$H = H_0 + V. \tag{7.204a}$$

The second-order memory function is given by

$$F_2(t) = -(QL^2A, e^{-iQLQt}QL^2A)(LA, LA)^{-1}. \tag{7.204b}$$

The approximation consists of omitting the Q operator in the time evolution. That is, we write (7.204b) as

$$F_2(t) = -(QL^2A,\, e^{-iL_0t}QL^2A)(LA,\, LA)^{-1}, \tag{7.205}$$

where $L_0 = i[H_0, .]$. The static quantities are calculated using H, but the time dependence is calculated using H_0. For more details, see Pires [8].

The memory function formalism is very convenient to study systems where the excitations cannot be considered as quasiparticles [11], as is the case in the neighborhood of a quantum critical point.

7.12 Hydrodynamic fluctuations

In practice, the computational difficulties involved in calculating correlation functions are non-trivial [6]. The part of these functions which varies slowly in space and time and reflects the hydrodynamic limit is the most difficult to compute. I will consider in the following the simplest example of a transport process: spin diffusion in a Heisenberg ferromagnet in the paramagnetic phase. In this case, the only hydrodynamic equation is a diffusion equation for the spin magnetization. The diffusion of magnetization takes place via spin-flip interaction between neighbor sites. If at a given point in space, the spin has the same probability to point up or down along a chosen direction of spin quantization, the magnetization vanishes at this point. However, if there is an imbalance between the density of spins pointing in the two directions, we have a magnetization proportional to the densities' difference. Let us represent by $m(\vec{r},\, t)$ the magnetization in the direction of quantization. Since the total spin is a constant of motion, we can assume that the total magnetization is conserved. That is

$$\frac{d}{dt}\int m(\vec{r},\, t)d\vec{r} = 0. \tag{7.206}$$

The conservation law implies a continuity equation for the magnetization:

$$\frac{\partial}{\partial t}m(\vec{r},\, t) + \vec{\nabla}\cdot\vec{j}\,(\vec{r},\, t) = 0, \tag{7.207}$$

where $\vec{j}\,(\vec{r},\, t)$ is the magnetization current. In the presence of an external magnetic field $\vec{B}(\vec{r})$, we can write the following phenomenological expression:

$$\langle\vec{j}\,(\vec{r},\, t)\rangle = -D\vec{\nabla}.\,\langle m(\vec{r},\, t)\rangle + \sigma\vec{\nabla}\cdot\vec{B}(\vec{r}). \tag{7.208}$$

The last term is responsible for the induced spin current due to the field gradient studied in section 7.6. Equation (7.208) is valid only if all the properties of the system vary slowly in space and time. The transport coefficient D is called the spin diffusion coefficient and is positive. Equation (7.207) is microscopically exact. However, equation (7.208) can be valid only on average, which is why I have put brackets around it. That indicates a non-equilibrium average since in thermal equilibrium $m(\vec{r},\, t)$ is independent of position and time, and the magnetization current vanishes.

Suppose a system in which B is time-independent but a slow function of $\vec{r}$. Eventually, the system reaches thermal equilibrium in which case $\langle m(\vec{r})\rangle = \chi B(\vec{r})$.

Under this condition of equilibrium, the transport current should vanish, and we get $\sigma = \chi D$ (a well-known Einstein relation). In the following, I will take $B = 0$.

Taking equation (7.208) into (7.207) we obtain the diffusion equation:

$$\frac{\partial}{\partial t}\langle m(\vec{r}, t)\rangle - \nabla^2\langle m(\vec{r}, t)\rangle = 0. \tag{7.209}$$

To eliminate boundary conditions, we will consider an infinitely extended system. Supposing that the system has been disturbed from thermodynamical equilibrium at a time $t = 0$, we can take the Fourier–Laplace transform, and obtain

$$\langle m(\vec{k}, z)\rangle = \int d\vec{r} e^{-i\vec{k}\cdot\vec{r}} \int_0^\infty dt e^{izt}\langle m(\vec{r}, t)\rangle. \tag{7.210}$$

Using (7.209) and (7.210) we find

$$\langle m(\vec{k}, z)\rangle = \frac{i}{z + iDk^2}\langle m(\vec{k}, t = 0)\rangle. \tag{7.211}$$

The diffusion process is reflected in a pole on the negative imaginary axis, at $z = -iDk^2$. From equation (7.211) we get

$$\langle m(\vec{k}, t)\rangle = \langle m(\vec{k}, t = 0)\rangle e^{-Dk^2 t}. \tag{7.212}$$

Equation (7.212) shows that a 'hydrodynamic mode' is a collective fluctuation that is exponentially damped for large wavelengths, with a lifetime $\tau = 1/Dk^2$. Since the magnetization is conserved, a local excess of this quantity cannot disappear locally but only relax by spreading slowly over the entire system.

To proceed, let us assume that equations (7.208) and (7.209) are valid even if we omit the average signs. That means that we can use the equations for operators. We multiply then (7.209) from the right by m (0,0) and then do an equilibrium average. We have then

$$\left(\frac{\partial}{\partial t} - D\nabla^2\right)S(\vec{r}, t) = 0. \tag{7.213}$$

That means that spontaneous fluctuations described by $S(\vec{r}, t)$ relax back to thermal equilibrium according to the same diffusion equation as do induced non-equilibrium fluctuations described by $\langle m\rangle_{\text{non-eq}}$. Equation (7.213) leads to

$$\tilde{S}(\vec{k}, z) = \frac{i}{z + iDk^2}S(\vec{k}, t = 0). \tag{7.214}$$

Having in mind that equation (7.214) holds only for Im $z > 0$ we can write

$$S(\vec{k}, \omega) = \frac{2Dk^2}{\omega^2 + (Dk^2)^2}S(\vec{k}, t = 0). \tag{7.215}$$

Equation (7.215) can be derived rigorously using the fluctuation-dissipation theorem [6] and is valid only for small k and ω. At short distances (large k), effects due to the lattice structure, the exact spin dynamics, and the residues of short-range order in the system will falsify the above procedure. The effect that the correlation function is extremely long-lived at small wavevectors has been named 'kinematic' slowing down by de Gennes. This reflects the fact that longer and longer wavelength fluctuations involve larger and larger aggregates of magnetization that take longer and longer times to disperse at a given rate of diffusion.

Hence, at temperatures sufficiently above the ordering temperature where propagating spin waves are not significant, the dynamic of a ferromagnet is governed by a diffusive process in the limits $\vec{r}, t \to \infty$. On the other side, in frustrated magnets, diffusion results from the macroscopic number of zero-energy modes in the instantaneous ground state, causing the system to wander around the degenerate manifold [12].

It is well known that in a gas, diffusion occurs due to collisions between particles and can be studied using the Boltzmann transport equation [13].

7.13 A brief discussion about experimental techniques

Thermodynamic properties of a system (such as specific heat and magnetic susceptibility) can easily be obtained from the partition function (as is shown in all books of statistical mechanics). However, these properties contain information neither on spatial structures nor on dynamical features. On the other side, correlation functions can be used to probe both static and dynamic features of a system. For a survey of how experimental techniques are related to correlation functions, see chapter 7 of Altland and Simons [14].

What is the physical meaning of the correlation function? Suppose the thermal average of a field $\langle\phi(0)\rangle$ vanishes, what happens, for instance, in the disordered phase. However, the average of the square amplitude $(\phi(0))^2$ is certainly non-vanishing. Now let us consider $\langle\phi(0)\phi(\vec{r})\rangle$. For asymptotically large values of $\vec{r}$, it is likely that the amplitudes fluctuate independently of each other and $\langle\phi(0)\phi(\vec{r})\rangle \to 0$ as $|\vec{r}| \to \infty$. However, locally the field amplitudes are correlated over a specific range in space. The correlation function measures how the field $\phi(\vec{r}, t)$ is correlated at two different points in space and time. As was said before, it describes the spontaneous fluctuations which arise from the thermal motion of the particles of the system. As we saw in chapter 5, the correlation function, in the disordered phase, decays exponentially at large distances

$$\langle\phi(0)\phi(\vec{r})\rangle \sim \exp(-|\vec{r}|/\xi), \tag{7.216}$$

and this equation defines the correlation length ξ. In the ordered phase, we have

$$\langle\phi(0)\phi(\vec{r})\rangle = \langle\phi(0)\rangle\langle\phi(\vec{r})\rangle = m^2, \tag{7.217}$$

where $\vec{m} \equiv \langle\phi(\vec{r})\rangle$. Equation (7.217) suggests that ξ diverges as $T \to T_c^+$.

One experimental technique of particular interest here is the inelastic neutron scattering experiment, where a crystal is bombarded by incident neutrons that interact weakly with the system. In a single scattering event, the incident neutron is scattered from a state $|\vec{k}\rangle$ to a state $|\vec{k}'\rangle$, and the state of the solid is changed from $|i\rangle$

of energy E_i to $|f\rangle$ of energy E_f. If $\vec{K} = \vec{k} - \vec{k}'$ is the change in wavevector of the incident neutron and $\omega = E_i - E_f$ the energy transfer to the target, we can show that the inelastic differential scattering cross-section per unit solid angle per unit energy range is proportional to the correlation function $S(\vec{K}, \omega)$ [15].

We can calculate correlation functions analytically only for noninteracting particles, as was done in section 7.5. For an interacting system, we should resort to calculating the relaxation function (using, for instance, the memory function formalism) or the powerful Green function technique (as discussed in section 7.9, and as will be presented in chapter 8). These functions are related to the correlation function by the fluctuation-dissipation theorem.

7.14 Exercises

7.1. In equation (7.18) consider a perturbation of the form

$$f(t) = \begin{cases} Fe^{\eta t} & t < 0 \\ 0 & t > 0 \end{cases},$$

then we can write

$$\Delta B(t) = \int_{-\infty}^{0} K(t - t')e^{\eta t'} F dt'.$$

Show that

$$\Delta B(z) = \frac{F}{\pi i}\int_{-\infty}^{\infty} \frac{\chi''(\omega)}{\omega(\omega - z)} d\omega.$$

7.2. The rate of energy transferred to a system (energy dissipation rate) can be calculated from

$$P = \left(\frac{d\langle H\rangle}{dt}\right)_{\text{average}}.$$

Using $V(t) = V_0 e^{-i\omega t}$ show that

$$P = \frac{\omega}{2}|V_0|^2 \chi''(\omega).$$

7.3. Using the linear spin-wave formalism find $\langle S_q^\alpha(t) S_{-q}^\alpha\rangle$ $(\alpha = x, z)$ for the cubic Heisenberg ferromagnet lattice. Calculate the time Fourier transform.

7.4. Using equation (7.4), assume that initially, the magnetization is non-zero only at $\vec{r} = 0$. That is $\langle m(\vec{r}, t = 0)\rangle = m\delta(\vec{r})$. Show that

$$\langle m(\vec{r}, t)\rangle = m(4\pi DT)^{-3/2}\exp(-r^2/Dt),$$

which displays the characteristic Gaussian spreading of a random walk process.

7.5. Show that at very high temperatures, in the limit $t \to \infty$, we have for the correlation function

$$\langle S_i^z(t) S_i^z \rangle \propto t^{-d/2},$$

where d is the dimension of the system.

References

[1] Shankar R 1994 *Principles of Quantum Mechanics* (New York: Plenum Press)
[2] Allen P B 2006 *Comtemp. Concepts Conden. Matter Sci.* **2** 165–218
[3] Sentef M, Kollar M and Kampf A P 2007 Spin transport in Heisenberg antiferromagnets in two and three dimensions *Phys. Rev.* B **75** 214403
[4] Nolting W and Ramakanth A 2009 *Quantum Theory of Magnetism* (Berlin: Springer)
[5] Ziman J M 1969 *Elements of Advanced Quantum Theory* (Cambridge: Cambridge University Press)
[6] Foster D 1975 *Hydrodynamics Fluctuations, Broken Symmetry, and Correlation Functions* (Reading, MA: Benjamin)
[7] Florencio J and Alcantara Bonfin O F 2020 Recent advances in the calculation of dynamical correlation functions arXiv:2005.14222
[8] Pires A S T 1988 The memory function formalism in the study of the dynamics of a many-body system *Helv. Phys. Acta* **61** 988–1006
[9] Lee M H 1982 Solutions of the generalized Langevin equation by a method of recurrence relations *Phys. Rev.* B **26** 2547–51
[10] Mori H 1965 A continued-fraction representation of the time-correlation functions *Prog. Theor. Phys.* **34** 399–416
[11] Lucas A and Sachdev S 2015 Memory matrix theory of magnetotransport in strange metals *Phys. Rev.* B **91** 195122
[12] Saha P, Zhang D, Lee S H and Chern G W 2021 Spin dynamics of the antiferromagnetic Heisenberg model on Kagome bilayers arXiv:1904.05863
[13] Sachdev S 2011 *Quantum Phase Transitions* (Cambridge: Cambridge University Press)
[14] Altland A and Simons B 2006 *Condensed Matter Field Theory* (Cambridge: Cambridge University Press)
[15] Kittel C 1963 *Quantum Theory of Solids* (New York: Wiley)

IOP Publishing

Theoretical Tools for Spin Models in Magnetic Systems

Antonio Sergio Teixeira Pires

Chapter 8

Perturbation theory

In most of the books using quantum field theory, perturbation theory is done for fermions. The use of this theory to the case of a system of interacting bosons at temperatures below the temperature of the Bose–Einstein (BE) condensation presents some difficulties [1]. The BE condensation occurs when the energy of the boson with the lowest energy in the gas becomes equal to the chemical potential. In this chapter, I will be interested in magnons. Magnons form a Bose gas with a chemical potential zero, and the energy of magnons is always greater than zero. Only in particular conditions is the chemical potential of magnons raised to lead to a BE condensation. Therefore, we can use the standard perturbation theory to study magnons' interactions. As was mentioned in chapter 7, Green functions contain the essential physical information of a system, such as the ground-state energy and other thermodynamic functions, the energy and lifetime of excited states, and the linear response to external perturbations [2]. Unfortunately, it is difficult to calculate the exact Green functions, and we, therefore, need to use perturbation theory, which I present in this chapter.

8.1 The interaction representation

In chapter 7, we studied a system subjected to an external perturbation. We added an external field to determine the response of the system to this field and calculated all averages using H_0. In several problems, we can use this procedure considering that H_0 is, in general, a harmonic Hamiltonian (sometimes an effective harmonic Hamiltonian). However, it can happen that we cannot do that, and we have to treat H_0 using perturbation theory. To proceed, I will rename H_0 as H and write

$$H = H_0 + H_I, \tag{8.1}$$

where here H_0 can be solved exactly, and H_I is now an intrinsic perturbation (and not an interaction with an external field); for example, the four-term operators in equation (3.43). There are a lot of good books treating the subject, for instance,

doi:10.1088/978-0-7503-3879-0ch8

Abrikosov *et al* [1], Fetter and Waleck [2], and Mahan [3], and here I will present only the general idea.

The Schrödinger equation for a state vector $|\psi_s(t)\rangle$ is (setting $\hbar = 1$)

$$i\frac{\partial}{\partial t}|\psi_s(t)\rangle = H|\psi_s(t)\rangle, \tag{8.2}$$

where there is no explicit time dependence in H. A formal solution to (8.2) is easily obtained as

$$|\psi_s(t)\rangle = e^{-iH(t-t_0)}|\psi_s(t_0)\rangle. \tag{8.3}$$

Given the solution at the time t_0, the transformation (8.3) generates the solution at time t. We define the interaction state vector as follows [2]

$$|\psi_I(t)\rangle \equiv e^{iH_0t}|\psi_s(t)\rangle. \tag{8.4}$$

The equation of motion of this state vector is

$$\begin{aligned} i\frac{\partial}{\partial t}|\psi_I(t)\rangle &= -H_0e^{iH_0t}|\psi_s(t)\rangle + e^{iH_0t}i\frac{\partial}{\partial t}|\psi_s(t)\rangle \\ &= e^{iH_0t}(-H_0 + H_0 + H_I)e^{-iH_0t}|\psi_I(t)\rangle, \end{aligned} \tag{8.5}$$

which can be written as

$$i\frac{\partial}{\partial t}|\psi_I(t)\rangle = H_I(t)|\psi_I(t)\rangle, \tag{8.6}$$

where

$$H_I(t) \equiv e^{iH_0t}H_Ie^{-iH_0t}. \tag{8.7}$$

In most of the problems, H_0 does not commute with H_I, therefore, the proper order of these operators is essential.

We may write an arbitrary matrix element of an operator A_s in the Schrödinger representation as

$$\langle\psi'_s(t)|A_s|\psi_s(t)\rangle = \langle\psi'_I(t)|A_I(t)|\psi_I(t)\rangle, \tag{8.8}$$

where

$$A_I(t) = e^{iH_0t}A_se^{-iH_0t}. \tag{8.9}$$

We write for $|\psi_I(t)\rangle$

$$|\psi_I(t)\rangle = U(t, t_0)|\psi_I(t_0)\rangle, \tag{8.10}$$

where $U(t, t_0)$ is a unitary operator. Note that $U(t_0, t_0) = 1$.

From (8.7) and (8.10), we find that U satisfies the differential equation

$$i\frac{\partial}{\partial t}U(t, t_0) = H_I(t)U(t, t_0). \tag{8.11}$$

Integrating (8.11) from t_0 to t we obtain

$$U(t, t_0) = 1 - i\int_{t_0}^{t} dt_1 H_I(t_1)U(t_1, t_0). \tag{8.12}$$

Solving (8.12) by iteration, we find

$$\begin{aligned} U(t, t_0) = 1 &- i\int_{t_0}^{t} dt_1 H_I(t_1) + (-i)^2\int_{t_0}^{t} dt_1\int_{t_0}^{t_1} dt_2 H_I(t_1)H_I(t_2) \\ &+ \ldots + (-i)^n\int_{t_0}^{t} dt_1\int_{t_0}^{t_1} dt_2 \ldots \int_{t_0}^{t_{n-1}} dt_n H_I(t_1)\ldots H_I(t_n) + \ldots \end{aligned} \tag{8.13}$$

Let us consider the term $n = 2$, and change $t_1 \to t_2$, $t_2 \to t_1$. We obtain

$$I = \int_{t_0}^{t} dt_1\int_{t_0}^{t_1} dt_2 H_I(t_1)H_I(t_2) = \int_{t_0}^{t} dt_2\int_{t_0}^{t_2} dt_1 H_I(t_2)H_I(t_1). \tag{8.14}$$

Reversing the order of integration, we get

$$\begin{aligned} \int_{t_0}^{t} dt_2\int_{t_0}^{t_2} dt_1 &= \int_{t_0}^{t} dt_2\int_{t_0}^{t} dt_1 \theta(t_2 - t_1) = \int_{t_0}^{t} dt_1\int_{t_0}^{t} dt_2 \theta(t_2 - t_1) \\ &= \int_{t_0}^{t} dt_1\int_{t_1}^{t} dt_2. \end{aligned} \tag{8.15}$$

Hence

$$I = \int_{t_0}^{t} dt_1\int_{t_1}^{t} dt_2 H_I(t_2)H_I(t_1). \tag{8.16}$$

Then, we can write

$$I = \frac{1}{2}\int_{t_0}^{t} dt_1\int_{t_0}^{t_1} dt_2 H_I(t_1)H_I(t_2) + \frac{1}{2}\int_{t_0}^{t} dt_1\int_{t_1}^{t} dt_2 H_I(t_2)H_I(t_1), \tag{8.17}$$

where the first term on the right-hand side of (8.17) holds for $t_1 > t_2$, and the second for $t_1 < t_2$. $H_I(t_1)H_I(t_2)$ and $H_I(t_2)H_I(t_1)$ are time ordered in each term. Hence, we can write

$$I = \frac{1}{2}\int_{t_0}^{t} dt_1\int_{t_0}^{t_1} dt_2 TH_I(t_1)H_I(t_2) + \frac{1}{2}\int_{t_0}^{t} dt_1\int_{t_1}^{t} dt_2 TH_I(t_2)H_I(t_1), \tag{8.18}$$

where here T is the time-ordering operator. Thus

$$I = \frac{1}{2}\int_{t_0}^{t} dt_1\int_{t_0}^{t} dt_2 TH_I(t_1)H_I(t_2). \tag{8.19}$$

We get a similar expression to the nth term. The final result is

$$U(t, t_0) = 1 + \sum_{n=1}^{\infty}\frac{(-i)^n}{n!}\int_{t_0}^{t} dt_1 \ldots \int_{t_0}^{t} dt_n TH_I(t_1)\ldots H_I(t_n). \tag{8.20}$$

Or, written in a compact form

$$U(t, t_0) = T \exp\left[-i\int_{t_0}^{t} H_I(t_1)dt_1\right]. \tag{8.21}$$

8.2 Green functions

Now we look for a connection between state vectors in the interaction representation and in the Heisenberg representation, that is, a relation of the form

$$|\psi_I(t)\rangle = Q(t)|\psi_H\rangle, \tag{8.22}$$

where $Q(t)$ is a unitary operator. Using equation (8.10), we get

$$Q(t) = U(t, t_0)Q(t_0). \tag{8.23}$$

Equation (8.22) implies

$$Q(t) = U(t, \tau)R, \tag{8.24}$$

where τ is some instant of time, and R is a time-independent operator. To determine R, we use (8.4), and the relation between state vectors in the Schrödinger and Heisenberg representation $|\psi_s(t)\rangle = e^{-iHt}\,|\psi_H\rangle$ to write

$$e^{iH_0t} = U(t, \tau)Re^{iHt}. \tag{8.25}$$

Using $U(\tau, \tau) = 1$ we find

$$R = e^{iH_0\tau}e^{-iH\tau}. \tag{8.26}$$

We suppose that at a time $t = -\infty$ there is no interaction between particles, and it is then turned on infinitely slowly. Hence $R \to 1$ as $\tau \to -\infty$, and so

$$|\psi_I(t)\rangle = U(t)|\psi_H\rangle, \tag{8.27}$$

where $U(t) = U(t, -\infty)$. Using the property

$$U(t_2, t_1)U(t_1, t_0) = U(t_2, t_0), \text{ for } t_2 > t_1 > t_0,$$

we get

$$U(t_2, t_1) = U(t_2)U^{-1}(t_1). \tag{8.28}$$

According to (8.27) we conclude that the relation between operators in the interaction representation and the Heisenberg representation has the form

$$O_H(t) = U^{-1}(t)O_I(t)U(t). \tag{8.29}$$

Now, suppose we have an expression of the form

$$F = \left\langle\psi_H^0\middle|TA_H(t)B_H(t')C_H(t'')\ldots\middle|\psi_H^0\right\rangle,\; t > t' > t'' > \ldots \tag{8.30}$$

where $\left|\psi_H^0\right\rangle$ is the ground state. Using (8.29) we can write

$$\begin{aligned}&\left\langle\psi_H^0\middle|U^{-1}(t)A_I(t)U(t)U^{-1}(t')B_I(t')U(t')\ldots\middle|\psi_H^0\right\rangle\\ &=\left\langle\psi_H^0\middle|U^{-1}(\infty)U(\infty,\,t)A_I(t)U(t,\,t')B_I(t')\ldots\middle|\psi_H^0\right\rangle\\ &=\left\langle\psi_H^0\middle|U^{-1}(\infty)TA_I(t)B_I(t')C_I(t'')\ldots U(\infty)\middle|\psi_H^0\right\rangle.\end{aligned}\tag{8.31}$$

In the next step, we need to calculate $\left\langle\psi_H^0\right|U^{-1}(\infty) = [U(\infty)\left|\psi_H^0\right\rangle]^*$. Using (8.10) and (8.27), we get

$$\left|\psi_H^0\right\rangle = |\psi_I(-\infty)\rangle,\quad U(\infty)\left|\psi_H^0\right\rangle = |\psi_s(\infty)\rangle.$$

Thus, $U(\infty)\left|\psi_H^0\right\rangle$ is the function $|\psi_I(\infty)\rangle$ obtained from the ground-state function $|\psi_I(-\infty)\rangle$ by adiabatically turning on the interaction. It is known that the ground state of a system is nondegenerate. Therefore, it cannot make a transition into another state under the action of an infinitely slow perturbation. Thus, the function $|\psi_I(\infty)\rangle = U(\infty)\left|\psi_H^0\right\rangle$ can differ from $\left|\psi_H^0\right\rangle$ only by a phase factor, that is

$$U(\infty)\left|\psi_H^0\right\rangle = e^{ia}\left|\psi_H^0\right\rangle.\tag{8.32}$$

The final result is

$$F=\frac{\left\langle\psi_H^0\right|T(A_I(t)B_I(t')C_I(t'')\ldots U(\infty)\left|\psi_H^0\right\rangle}{\left\langle\psi_H^0\right|U(\infty)\left|\psi_H^0\right\rangle}.\tag{8.33}$$

The time ordered Green function at zero temperature is given by

$$G_{AB}(x,\,x') = -i\left\langle\psi_H^0\middle|TA(x)B(x')\middle|\psi_H^0\right\rangle,\tag{8.34}$$

where A and B are operators in the Heisenberg representation, and x denotes $\vec{r}$ and t and x' denotes $(\vec{r}',\,t')$. Using equation (8.33), we can write (8.34) as

$$G_{AB}(x,\,x') = -i\frac{\left\langle\psi_H^0\middle|TA_I(x)B_I(x')U(\infty)\middle|\psi_H^0\right\rangle}{\left\langle\psi_H^0\middle|U(\infty)\middle|\psi_H^0\right\rangle}.\tag{8.35}$$

It is usual to omit the lower index 'I' and take $|0\rangle = \left|\psi_H^0\right\rangle$. Equation (8.35) is then written as

$$G_{AB}(x,\,x') = -i\frac{\langle 0|TA(x)B(x')U(\infty)|0\rangle}{\langle 0|U(\infty)|0\rangle}.\tag{8.36}$$

Using equation (8.20), equation (8.36) takes the form

$$\begin{aligned}G_{AB}(x,\,x') = -\frac{i}{\langle 0|U(\infty)|0\rangle}\sum_{n=0}^{\infty}\frac{(-i)^n}{n!}\int_{-\infty}^{\infty}\ldots\int_{-\infty}^{\infty}dt_1\ldots dt_n\\ \langle 0|TA(x)B(x')H(t_1)\ldots H(t_n)|0\rangle.\end{aligned}\tag{8.37}$$

8.3 Wick's theorem

To use equation (8.37), we have to calculate a vacuum expectation value of a time ordered string of operators, which is quite complicated. On the other hand, a vacuum expectation value of a normal ordered string of operators is easy to evaluate. Normal ordering places annihilation operators on the right and creation operators on the left. Hence, the result is identically zero. To relate the two quantities, we demonstrate the so-called Wick's theorem, following Lurie [4].

We can write an operator $A(x)$ of our interest as $A(x) = A^{(+)}(x) + A^{(-)}(x)$, where $A^{(+)}$ contains annihilation operators and $A^{(-)}$ creation operators. Thus, we have $A^{(-)}\,|0\rangle = 0$, and $\langle 0|\, A^{(+)} = 0$. We start with the trivial relation

$$TA(x_1) = :A(x_1): \tag{8.38}$$

where :: indicates the normal ordering. We multiply (8.38) on the right by an arbitrary operator $B(x_2)$, with $t_2 < t_1$ and get

$$(TA(x_1))B(x_2) = TA(x_1)B(x_2) = :A(x_1):B(x_2). \tag{8.39}$$

To absorb $B(x_2)$ into the normal product, we write

$$\begin{aligned} :A(x_1)&: B(x_2) = A(x_1)B^{(+)}(x_2) + A(x_1)B^{(-)}(x_2) \\ &= A(x_1)B^{(+)}(x_2) + A^{(-)}(x_1)B^{(-)}(x_2) + A^{(+)}(x_1)B^{(-)}(x_2). \end{aligned} \tag{8.40}$$

All terms on the right-hand side of equation (8.40) except $A^{(+)}B^{(-)}$ are in a normal ordered form. The latter term may be normal ordered by commuting $B^{(-)}$ through to the left using the relation

$$A^{(+)}(x_1)B^{(-)}(x_2) = A^{(-)}(x_2)B^{(+)}(x_1) + [A^{(+)}(x_1),\, B^{(-)}(x_2)]. \tag{8.41}$$

A commutator is a complex number, hence, we have for $t_2 < t_1$

$$\begin{aligned} [A^{(+)}(x_1),\, B^{(-)}(x_2)] &= \langle 0|[A^{(+)}(x_1),\, B^{(-)}(x_2)]|0\rangle \\ &= \langle 0|A^{(+)}(x_1)B^{(-)}(x_2)|0\rangle \\ &= \langle 0|A(x_1)B(x_2)|0\rangle \\ &= \langle 0|TA(x_1)B(x_2)|0\rangle. \end{aligned} \tag{8.42}$$

Putting together the above results, we obtain

$$TA(x_1)B(x_2) = :A(x_1)B(x_2): + \langle 0|TA(x_1)B(x_2)|0\rangle, \tag{8.43}$$

where $t_2 < t_1$. This restriction can be removed by observing that $A(x_1)$ and $B(x_2)$ can now be interchanged on both sides of (8.43). By the definition of the normal ordered and time ordered products, equation (8.43) remains invariant under permutation, and so it is valid in general.

The above derivation can be generalized to higher-order products. For three operators, we multiply (8.43) on the right by $C(x_3)$, with t_1, $t_2 > t_3$:

$$\begin{aligned} TA(x_1)B(x_2)C(x_3) = (TA(x_1)B(x_2))C(x_3) &= :A(x_1)B(x_2):C(x_3) \\ &+ \langle 0|TA(x_1)B(x_2)|0\rangle C(x_3). \end{aligned} \tag{8.44}$$

In the same way as before, $C(x_3)$ can be absorbed into the normal product

$$:A(x_1)B(x_2):C(x_3) = :A(x_1)B(x_2):C^{(+)}(x_3) + :A(x_1)B(x_2):C^{(-)}(x_3). \tag{8.45}$$

The first term on the right-hand side of (8.45) is already in normal ordered form. We can normal order the last term by successively commuting $C^{(-)}(x_3)$ through to the left using (8.42). We obtain

$$\begin{aligned} &:A(x_1)B(x_2):\ C^{(-)}(x_3) = :A(x_1)B(x_2)C^{(-)}(x_3): \\ &+ A(x_1)\langle 0|TB(x_2)C(x_3)|0\rangle \\ &+ B(x_2)\langle 0|TA(x_1)C(x_3)|0\rangle. \end{aligned} \tag{8.46}$$

From (8.44), (8.45) and (8.46), we find

$$\begin{aligned} &TA(x_1)B(x_2)C(x_3) = :A(x_1)B(x_2)C(x_3): + C(x_3)\langle 0|TA(x_1)B(x_2)|0\rangle \\ &+ B(x_2)\langle 0|TA(x_1)C(x_3)|0\rangle + A(x_1)\langle 0|TB(x_2)C(x_3)|0\rangle, \end{aligned} \tag{8.47}$$

for t_1, $t_2 > t_3$. We can lift the latter restriction in the same way as we did before by noting that (8.47) remains invariant under any permutation of $A(x_1)$, $B(x_2)$ and $C(x_3)$, as long as the same permutation is applied to both sides of the equation.

Now, we introduce the following compact notation referred to as the contraction of A with B

$$\underbracket{A(x_1)B(x_2)} = \langle 0|TA(x_1)B(x_2)|0\rangle. \tag{8.48}$$

Using this notation, and following the same procedure as before we find that the decomposition formula for a four-fold product is given by

$$\begin{aligned} &TA(x_1)B(x_2)C(x_3)D(x_4) =: A(x_1)B(x_2)C(x_3)D(x_4): \\ &+ :\underbracket{A(x_1)B(x_2)}C(x_3)D(x_4): + \text{all other singly contracted normal products} \\ &+ :\underbracket{A(x_1)B(x_2)}\,\underbracket{C(x_3)D(x_4)}: + \text{all other doubly contracted normal products.} \end{aligned} \tag{8.49}$$

In equation (8.49), we also have six singly contracted normal products:

$$:\underbracket{AB}CD:,\ :\underbracket{ABC}D:,\ :\underbracket{ABCD}:,\ :A\underbracket{BC}D:,\ :A\underbracket{BCD}:,\ :AB\underbracket{CD}:,$$

and three doubly contracted products

$$:\underbracket{AB}\,\underbracket{CD}:,\quad \underbracket{A\overline{B\,C}}\overline{D}\ ,\ :\underbracket{A\,\overbracket{BC}\,D}:$$

The above expression can be generalized for higher-order products as:

$$T(ABC\ldots Z) = :ABC\ldots Z: + \text{all possible contractions of } ABC\ldots Z. \tag{8.50}$$

This result is called Wick's theorem. When we calculate the vacuum expectation value of (8.50), the only terms that survive are the ones involving all operators' contractions. Using Wick's theorem, we can evaluate the exact Green function (8.37) as a perturbation expansion involving only free-field Green functions.

8.4 Feynman diagrams

Only for a pedagogical purpose, I will present a simple example, which is similar to what is discussed in several textbooks. In general, the authors treat fermions, but here I will consider bosons and write for the interaction Hamiltonian

$$H_{\text{int}} = \frac{1}{2}\int \phi^{+}(\vec{r}_1)\phi^{+}(\vec{r}_2)U(\vec{r}_1 - \vec{r}_2)\phi(\vec{r}_2)\phi(\vec{r}_1)d\vec{r}_1 d\vec{r}_2, \tag{8.51}$$

where ϕ are boson fields and ϕ^{+} is the adjoint field, and following Abrikosov [1], I introduce the function $V(x_1 - x_2) = U(\vec{r}_1 - \vec{r}_2)\delta(t_1 - t_2)$. The first term in (8.37) is the Green function for noninteracting particles. The second term in the integrand is

$$\langle 0|\; T\phi(x)\phi^{+}(x')\phi^{+}(x_1)\phi^{+}(x_2)\phi(x_2)\phi(x_1)|0\rangle\, V(x_1 - x_2). \tag{8.52}$$

Using Wick's theorem, we can write the matrix elements in (8.52) as

$$\begin{aligned}
&\langle 0|T\phi(x)\phi^{+}(x_1)|0\rangle\langle 0|T\phi^{+}(x_2)\phi(x_2)|0\rangle\langle 0|T\phi(x_1)\phi^{+}(x')|0\rangle\\
&+\langle 0|T\phi(x)\phi^{+}(x_1)|0\rangle\langle 0|T\phi^{+}(x_2)\phi(x_1)|0\rangle\langle 0|T\phi(x_2)\phi^{+}(x')|0\rangle\\
&+\langle 0|T\phi(x)\phi^{+}(x_2)|0\rangle\langle 0|T\phi^{+}(x_1)\phi(x_1)|0\rangle\langle 0|T\phi(x_2)\phi^{+}(x')|0\rangle\\
&+\langle 0|T\phi(x)\phi^{+}(x_2)|0\rangle\langle 0|T\phi^{+}(x_1)\phi(x_2)|0\rangle\langle 0|T\phi(x_1)\phi^{+}(x')|0\rangle\\
&+\langle 0|T\phi(x)\phi^{+}(x')|0\rangle\langle 0|T\phi^{+}(x_1)\phi(x_1)|0\rangle\langle 0|T\phi^{+}(x_2)\phi(x_2)|0\rangle\\
&+\langle 0|T\phi(x)\phi^{+}(x')|0\rangle\langle 0|T\phi^{+}(x_1)\phi(x_2)|0\rangle\langle 0|T\phi^{+}(x_2)\phi(x_1)|0\rangle.
\end{aligned} \tag{8.53}$$

Using equation (8.34), equation (8.53) becomes

$$\begin{aligned}
&iG^{0}(x, x_1)G^{0}(x_2, x_2)G^{0}(x_1, x') + iG^{0}(x, x_1)G^{0}(x_1, x_2)G^{0}(x_2, x')\\
&+ iG^{(0)}(x, x_2)G^{(0)}(x_1, x_1)G^{(0)}(x_2, x') + iG^{(0)}(x, x_2)G^{(0)}(x_2, x_1)G^{(0)}(x_1, x')\\
&+ iG^{(0)}(x, x')G^{(0)}(x_1, x_1)G^{(0)}(x_2, x_2) + iG^{(0)}(x, x')G^{(0)}(x_2, x_1)G^{(0)}(x_1, x_2).
\end{aligned} \tag{8.54}$$

We have thus a sum of terms, each of which contains three noninteracting Green functions. It is possible to associate with each term a diagram following the rule: We draw points to represent the set of space–time coordinates on which the operator ϕ depends. Next, we connect by a solid line the points appearing as arguments in a function $G^{(0)}$ and by a dashed line the points appearing in $V(x_1 - x_2)$. This procedure gives the six diagrams for $G^{(1)}$ as shown in figure 8.1, where x and x' are external and x_1, x_2 internal points. Then, we integrate over the coordinates of the internal points. We can use the same procedure for higher-order terms and other forms of interactions. The obtained diagrams are called *Feynman diagrams*.

We can divide the Feynman diagrams into two groups: connected diagrams and disconnected diagrams. In a connected diagram, all points are connected by lines to the external points x and x'. In figure 8.1 the diagrams in the first line are connected.

Otherwise, the diagrams are disconnected, such as the diagrams in the second line. It can be shown [1] that the contribution of all disconnected diagrams cancels the term $\langle 0|U(\infty)|0\rangle$ in the denominator of equation (8.37). Another simplification is to omit the factor $1/n!$ and consider only pairing that leads to topologically non-equivalent diagrams, i.e., diagrams that cannot be obtained from each other by permutation of the operators H_{int}.

When the times in both arguments of $G^{(0)}$ are the same, we interpret that as [1] $\lim_{\varepsilon\to 0} G(t, t+\varepsilon) \equiv \lim_{\varepsilon\to 0} G(-\varepsilon) = i\langle 0|\phi^{+}(\vec{r}_1)\phi(\vec{r}_2)|0\rangle$.

It is more convenient for uniform systems to work in the momentum space since the calculations turn out to be more manageable. To show how this is done, I will study the expression corresponding to the second diagram in the first line of figure 8.1.

$$i\int d^4x_1 d^4x_2 G^{(0)}(x-x_1)G^{(0)}(x_1-x_2)G^{(0)}(x_2-x')V(x_1-x_2). \tag{8.55}$$

We introduce the Fourier transforms

$$G^{(0)}(x_1-x_2) = \int \frac{d^4p}{(2\pi)^4} G^{(0)}(p)e^{ip(x_1-x_2)}, \tag{8.56}$$

$$V(x_1-x_2) = \int \frac{d^4q}{(2\pi)^4} V(q)e^{iq(x_1-x_2)}, \tag{8.57}$$

where $p = (\vec{p}, \omega)$, $q = (\vec{q}, \omega)$ and $p(x_1-x_2) = \vec{p}\,.\,(\vec{r}_1-\vec{r}_2) - \omega(t_1-t_2)$. Substituting (8.56) and (8.57) into (8.55) we get

$$\frac{i}{(2\pi)^{16}}\int d^4p_1 d^4p_2 d^4p_3 d^4q d^4x_1 d^4x_2 G^{(0)}(p_1)G^{(0)}(p_2)G^{(0)}(p_3)V(q)$$
$$\times e^{i[p_1(x-x_1)+p_2(x_1-x_2)+p_3(x_2-x')+q(x_1-x_2)]}. \tag{8.58}$$

Integrating with respect to x_1 and x_2 we obtain

$$\begin{aligned}&\frac{i}{(2\pi)^8}\int d^4p_1 d^4p_2 d^4p_3 d^4q G^{(0)}(p_1)G^{(0)}(p_2)G^{(0)}(p_3)V(q)\\ &\times \delta(p_1-p_2-q)\delta(p_2+q-p_3)e^{i(p_1x-p_3x')},\end{aligned} \tag{8.59}$$

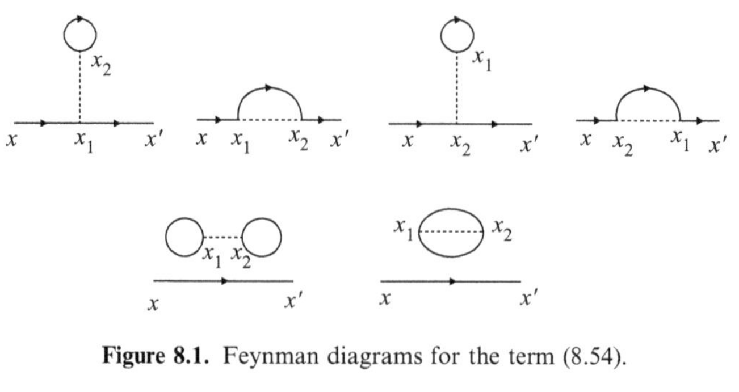

Figure 8.1. Feynman diagrams for the term (8.54).

which leads to

$$i\int d^4p_2 d^4q G^{(0)}(p)G^{(0)}(p_2)G^{(0)}(p')V(q)\delta(p - p_2 - q)\delta(p_2 + q - p'). \tag{8.60}$$

Integrating now with respect to p_2 we arrive at

$$G^{(1)}(p) = iG^{(0)}(p)\int \frac{d^4q}{(2\pi)^4}G^{(0)}(p - q)V(q)G^{(0)}(p). \tag{8.61}$$

The general rules for obtaining the expression corresponding to a diagram of order n are:

1. With each line, associate a four-momentum in such a way that the two external lines have the external momentum, and the momenta of the internal lines satisfy the conservation laws at each vertex.
2. With each solid line associate the Green function $G^{(0)}(p)$.
3. With each dashed line associate a quantity $V(q) = U(\vec{q})$.
4. Integrate over the n independent internal four-momenta.
5. Multiply the result by $i^n/(2\pi)^{4n}$.

In general, one calculates only lower-order diagrams, but we can add the contribution from all orders for a given diagram. For instance, consider the diagram

$\Sigma =$ 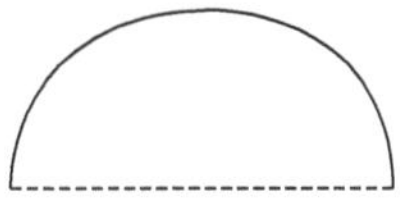

of figure 8.1. It will appear n times in the nth order, and then we can write

$$G = G^{(0)} + G^{(0)}\Sigma G^{(0)} + G^{(0)}\Sigma G^{(0)}\Sigma G^{(0)} + \dots \tag{8.62}$$

which can also be written as

$$G = G^{(0)} + G^{(0)}\Sigma G, \tag{8.63}$$

or

$$G = \frac{1}{(G^{(0)})^{-1} - \Sigma}. \tag{8.64}$$

We can distinguish between reducible and irreducible diagrams. A reducible diagram is one that may be separated in two by cutting just one line (figure 8.2).

Any part of a diagram which can be joined to the rest of the diagram by two $G^{(0)}$ lines is called a self-energy part. We can generalize equation (8.64) and consider Σ the sum of all different irreducible diagrams in the self-energy part of G. This generalized equation is called the *Dyson equation.* In appendix B, I will apply the formalism to treat a spin Hamiltonian.

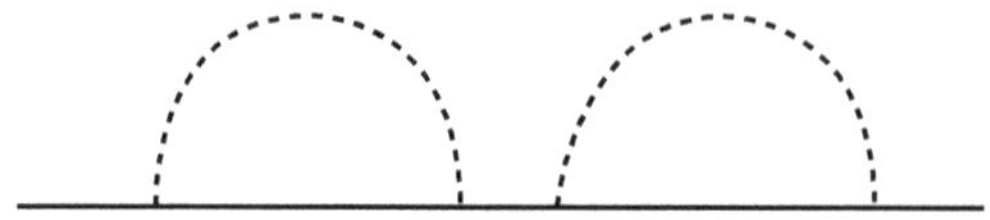

Figure 8.2. Reducible diagram. We see that we can separate the diagram into two parts by cutting the line between them.

8.5 Interpretation of the Green function

The Green function for free particles is generally given by

$$G^{(0)}(\vec{k}, \omega) = \frac{1}{\omega - E_0(k) + i\eta}. \tag{8.65}$$

Suppose that after performing a calculation taking into account the interaction between particles, we find

$$G(\vec{k}, \omega) = \frac{1}{\omega - E_0(\vec{k}) - \Sigma(\vec{k}, \omega)}. \tag{8.66}$$

We will write $\Sigma(\vec{k}, \omega) = \Sigma'(\vec{k}, \omega) + i\Sigma''(\vec{k}, \omega)$, where Σ' and Σ'' are the real and imaginary parts of Σ, respectively. In a first approximation, evaluating Σ at the energy where the peak occurs, we can write (8.66) as

$$G(\vec{k}, \omega) = \frac{A}{\omega - \varepsilon(\vec{k}) + i\Gamma}, \tag{8.67}$$

where $\varepsilon(\vec{k}) = E_0(\vec{k}) + \Sigma'(\vec{k}, \omega)$, and $\Gamma = -\Sigma''(\vec{k}, \omega)$. The Fourier transform of equation (8.67) is

$$G(\vec{k}, t) = \frac{1}{2\pi}\int_{-\infty}^{\infty} G(\vec{k}, \omega)e^{-i\omega t}d\omega. \tag{8.68}$$

For $t > 0$, the integral may be calculated using residues completing the contour in the lower half-plane. We find

$$G(\vec{k}, t) = iAe^{-i\varepsilon(\vec{k})t}e^{-\Gamma(\vec{k})t}, \tag{8.69}$$

which corresponds to an excitation with a finite lifetime, with a decaying rate $\Gamma(\vec{k})$. Hence, the original problem of a gas of free particles, becomes, when the interaction is taken into account, into a gas of free quasi-particles but with the energy shifted from the single-particle energy by the real part of the self-energy operator, and a finite lifetime. The appearance of a finite lifetime means that, in the presence of interactions, single-particle states no longer represent stable objects but rather tend to decay.

8.6 Two-particle Green function

The two-particle Green function (also called the two-body propagator) is defined by [5]

$$K(x_1, x_2; x_3, x_4) \equiv \langle 0|T\phi(x_1)\phi(x_2)\phi^+(x_3)\phi^+(x_4)|0\rangle. \tag{8.70}$$

Note that a two-particle Green function carries four sets of labels corresponding to each particle's propagation between two different points in space and time. We can show that for the case of noninteracting particles in the ground state, equation (8.70) reduces to [5]

$$K^{(0)}(x_1, x_2, x_3, x_4) = G^{(0)}(x_1, x_3)G^{(0)}(x_2, x_4) - G^{(0)}(x_1, x_4)G^{(0)}(x_2, x_3). \tag{8.71}$$

Using the technique described in section (8.4), we can write $G(x_1, x_2; x_3, x_4)$ in the form

$$K(x_1, x_2; x_3, x_4) = \frac{\langle 0|T\phi(x_1)\phi(x_2)\phi^+(x_3)\phi^+(x_4)U(\infty)|0\rangle}{\langle 0|U(\infty)|0\rangle}, \tag{8.72}$$

where, in the above equation, ϕ is an interaction picture field operator, and $|0\rangle$, as before, is the unperturbed vacuum state.

If we represent by $\Gamma(x_1, x_2; x_3, x_4)$ the sum of all irreducible interactions diagrams in $K(x_1, x_2; x_3, x_4)$, it can be shown that in momentum space we have [5]

$$\begin{aligned} K(k_1, k_2; k_3, k_4) &= G(k_1, k_3)G(k_2, k_4) - G(k_1, k_4)G(k_2, k_3) \\ &+ \frac{1}{2}\sum_{k_5,k_6} G(k_1)G(k_2)J(k_1, k_2; k_5, k_6)K(k_5, k_6, k_3, k_4). \end{aligned} \tag{8.73}$$

This equation is called the *Bethe–Salpeter equation.*

By space–time translation invariance $K(x_1, x_2, x_3, x_4)$ can depend only in differences of coordinates, say, $x = x_1 - x_2$, $x' = x_3 - x_4$ and $X - X' = \frac{1}{2}(x_1 + x_2) - \frac{1}{2}(x_3 + x_4)$. We introduce the Fourier transform $K(p, q, k)$ by

$$K(x_1, x_2, x_3, x_4) = \frac{1}{(2\pi)^8}\int d^4p d^4q d^4k e^{ik.(X-X')}e^{ipx}e^{-iq.x'}K(p, q, k). \tag{8.74}$$

The diagrammatic expansion of K is shown in figure 8.3 [6]. The lowest order contribution to this expansion consists of only two Green functions. The sum of the contributions involving all four operators containing interaction lines between the Green functions is shown next. The sum of all these contributions is represented by the diagram containing the shadowed surface. Some low order contributions to the expansion are shown in the second line. In this figure, all lines representing the single-particle Green functions represent the exact single-particle Green functions. All diagrams involving interaction between the two Green functions contain four external 'legs', i.e., the Green function connectors to the external field operators. If we remove these legs, we have what is called a *vertex*. A diagram will be called *reducible* if one can draw a line through it, which cuts no dashed lines at all and each of two solid lines only once. If such a line cannot be drawn, the diagram is called *irreducible*.

We can group all irreducible contributions to the vertex into a structure called an *irreducible vertex*, shown in figure 8.3 as a lightly hatched area. We denote the full

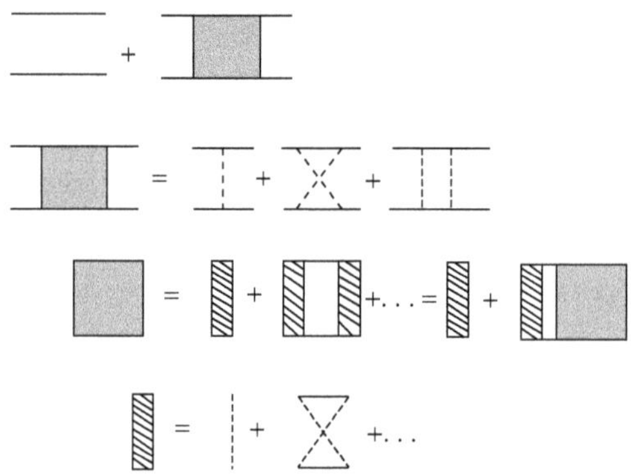

Figure 8.3. Feynman diagrams of the two-particle Green function. External lines represent the single-particle Green function.

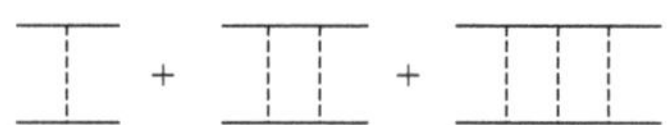

Figure 8.4. Ladder approximation for the two-particle Green function.

vertex by the symbol Γ and the irreducible vertex by V. In practice, the Bethe–Salpeter equation can be treated only in the so-called ladder approximation. That is, we consider only the diagrams shown in figure 8.4. Also, we replace the dressed propagators G by their lowest order values $G^{(0)}$.

8.7 Antiferromagnet

In this section, I will comment on how the formalism developed in this chapter can be used to study the Heisenberg antiferromagnet discussed in chapter 3. The Hamiltonian with the perturbation term was given in equation (3.43). I will only sketch the general idea. A detailed study can be found in Canali and Girvin [7] and Sentef [8].

For an antiferromagnet, the basic propagators are

$$\begin{aligned} G_{\alpha\alpha}(\vec{k}, t - t') &= -i\langle 0|T\alpha_k(t)\alpha_k^{\dagger}(t')|0\rangle, \\ G_{\beta\beta}(\vec{k}, t - t') &= -i\langle 0|T\beta_k^{\dagger}(t)\beta_k(t')|0\rangle, \\ G_{\alpha\beta}(\vec{k}, t - t') &= -i\langle 0|T\alpha_k(t)\beta_k(t')|0\rangle, \\ G_{\beta\alpha}(\vec{k}, t - t') &= -i\langle 0|T\beta_k^{\dagger}(t)\alpha_k^{\dagger}(t')|0\rangle. \end{aligned} \tag{8.75}$$

The Fourier transform is defined by

$$G_{\mu\nu}(\vec{k}, \omega) = \int_{-\infty}^{\infty} d(t - t')e^{i\omega(t-t')}G_{\mu\nu}(\vec{k}, t - t'), \tag{8.76}$$

where μ, ν can take the values α, β. The free particle propagators are easily found to be

$$G^{(0)}_{\alpha\alpha}(\vec{k}, \omega) = \frac{1}{\omega - \omega_k + i\eta}, \quad G^{(0)}_{\beta\beta}(\vec{k}, \omega) = -\frac{1}{\omega + \omega_k - i\eta}, \tag{8.77}$$

and $G^{(0)}_{\alpha\beta} = G^{(0)}_{\beta\alpha} = 0$. Note that ω_k is the energy associated with the α magnon and $-\omega_k$ to the β magnon. That is the reason for the definitions in equation (8.75).

As we saw in section (8.4), $G_{\mu\nu}$ satisfy the Dyson equation (8.63)

$$G_{\mu\nu}(\vec{k}, \omega) = G^{(0)}_{\mu\nu} + \sum_{\gamma,\delta} G^{(0)}_{\mu\gamma}(\vec{k}, \omega)\Sigma_{\gamma\delta}(\vec{k}, \omega)G_{\delta\nu}(\vec{k}, \omega), \tag{8.78}$$

where the magnon self-energy $\Sigma_{\gamma\delta}(\vec{k}, \omega)$ can be expressed in terms of a diagrammatic expansion.

In several problems leading with magnetic systems (for instance, in the calculation of spin currents), we are interested in the computation of the two-particle Green function

$$\Pi_{kq}(t - t') = -i\langle 0|T\alpha_k(t)\beta_k(t)\alpha_q^{\dagger}(t')\beta_q^{\dagger}(t')|0\rangle. \tag{8.79}$$

We can write a general expression for $\Pi_{kq}(\omega)$ following the discussion of section (8.6) as

$$\Pi_{kq}(\omega) = i\int \frac{d\omega'}{2\pi} G_{\alpha\alpha}(\vec{k}, \omega + \omega')G_{\beta\beta}(\vec{k}, \omega')\Gamma_{kq}(\omega, \omega'), \tag{8.80}$$

where the vertex function $\Gamma_{kq}(\omega, \omega')$ satisfies the Bethe–Salpeter equation

$$\begin{aligned}\Gamma_{kq}(\omega, \omega') = \delta_{kq} - i\sum_{k_1}\int \frac{d\omega_1}{2\pi} V^{\alpha\beta}_{kk_1k_1k}(\omega', \omega_1)G_{\alpha\alpha}(\vec{k}_1, \omega + \omega_1)\\ G_{\beta\beta}(\vec{k}_1, \omega_1)\Gamma_{kq}(\omega, \omega').\end{aligned} \tag{8.81}$$

Here $V^{\alpha\beta}_{kk_1k_1k}$ is the sum of all the irreducible diagrams. The calculations are generally performed using the ladder approximation.

8.8 Finite temperature Green function

All the functions studied in the previous chapters were temperature dependent. However, in this chapter, to use a perturbation theory, we have examined the time ordered Green function (Feynman propagator) at zero temperature. In this section, I show how to extend the results to finite temperatures. We could define a temperature-dependent Green function as

$$G^T(\vec{k}, t_2 - t_1) = -i\langle T\{a_k(t_2)a_k^{\dagger}(t_1)\}\rangle \tag{8.82a}$$

where $\langle A\rangle = tr(\rho A)/tr\rho, \quad \rho = e^{-\beta H}$.

We can calculate (8.82a) for free particles; however, we cannot expand G^T in Feynman diagrams, as we did for the zero temperature Green function. Nevertheless, we can modify G^T into an imaginary time Green function as follows.

We define a temperature variable τ, with $0 \leqslant \tau \leqslant \beta$ (we set $\hbar = 1$) and the τ-dependent operator $A(\tau)$

$$A(\tau) = e^{H\tau} A e^{-H\tau}, \tag{8.82b}$$

which is formally obtained by the analytic continuation to imaginary time $\tau = it$ of the Heisenberg operator $A(t)$. Likewise

$$U(t) \to U_e(\tau) = e^{-H\tau}. \tag{8.83}$$

Thus, a quantum system in equilibrium at temperature T can be regarded as the imaginary time evolution of this system. Since the ensemble average is a trace in Fock space, the initial and final states are the same. In other words, the states obey periodic boundary conditions in the imaginary time coordinate τ. Hence, we will restrict the imaginary time variable to the range $0 \leqslant \tau \leqslant \beta$.

The temperature-dependent Green function, also called the Matsubara Green function, is defined as

$$G^M_{AB}(\tau, \tau') = -\langle T_\tau(A(\tau)B(\tau'))\rangle, \tag{8.84}$$

where $T_\tau A(\tau)B(\tau')$ is an imaginary time ordered product defined by

$$T_\tau\{A(\tau)B(\tau')\} = \theta(\tau - \tau')A(\tau)B(\tau') + \theta(\tau' - \tau)B(\tau')A(\tau). \tag{8.85}$$

The temperature-dependent Green function obeys periodic boundary conditions for boson operators, as will be considered here (in the case of fermions, it follows anti-periodic conditions). Since the imaginary time interval is finite, the propagators can be expanded in Fourier series as

$$G^M(\tau) = \frac{1}{\beta} \sum_{n=-\infty}^{\infty} G^M(\omega_n) e^{i\omega_n \tau}, \tag{8.86}$$

$$G^M(\omega_n) = \frac{1}{2} \int_{-\beta}^{\beta} d\tau G^M(\tau) e^{-i\omega_n \tau}. \tag{8.87}$$

The periodic boundary condition in τ is satisfied if the frequencies $\{\omega_n\}$ are given by $\omega_n = 2\pi n/\beta$ (for bosons). Since the temperature-dependent Green function is formally the analytic continuation of the time ordered Green function to imaginary time, restricted to the imaginary time interval $0 \leqslant \tau \leqslant \beta$, its imaginary time Fourier transform (or series), $G_F(\omega_n)$ can also be regarded as the analytic continuation from real frequency ω to the imaginary frequency axis, restricted to the values $i\omega_n$.

We can put a $\vec{k}$ dependence on G^M and write $G^M(\vec{k}, \omega_n)$. For free bosons, we have

$$\langle a_k^\dagger a_k \rangle = n_{k,} \quad \langle a_k a_k^\dagger \rangle = 1 + n_k. \tag{8.88}$$

From (8.82b) we get $a_k(\tau) = a_k e^{-\tau\omega_k}$. Hence, we can write

$$G^{M(0)}(\vec{k}, \tau_2 - \tau_1) = -\begin{cases} \langle a_k a_k^\dagger \rangle e^{-\omega_k(\tau_2-\tau_1)} & \tau_2 > \tau_1 \\ \langle a_k^\dagger a_k \rangle e^{-\omega_k(\tau_2-\tau_1)} & \tau_2 < \tau_1 \end{cases}. \tag{8.89}$$

Inserting (8.89) into (8.87) we get

$$G^{M(0)}(\vec{k}, \omega_n) = -\frac{1}{2}\left\{\int_{-\beta}^{0} e^{i\omega_n\tau} n_k e^{-\omega_k\tau} d\tau + \int_{0}^{\beta} e^{i\omega_n\tau}(1 + n_k)e^{-\omega_k\tau} d\tau\right\}, \tag{8.90}$$

which gives

$$G^{M(0)}(\vec{k}, \omega_n) = \frac{1}{i\omega_n - \omega_k}. \tag{8.91}$$

The Green function G^M may be expanded in a perturbation series similar to the one for G in the $T = 0$ case. For instance, the Dyson equation is given by

$$G^{M}(\vec{k}, \omega_n) = \frac{1}{i\omega_n - \omega_k - \Sigma(\vec{k}, \omega_n)}. \tag{8.92}$$

We can extend all the previous sections' calculations for zero temperatures to finite temperatures using the procedure described above. However, we can use a more straightforward approach as follows. First, we solve the problem at $T = 0$ and obtain G.

The temperature-dependent Green function can be obtained from the Green function G at $T = 0$ by doing the analytical continuation $\omega \to i\omega_n$ and replacing

$$\frac{1}{2\pi}\int d\omega \quad \text{by} \quad \frac{i}{\beta}\sum_n. \tag{8.93}$$

Then we do another analytical continuation to get ω real.

Some results that can be useful in the calculations of temperature-dependent Green functions are presented below.

$$\sum_n \frac{1}{i\omega_n - x} = \beta n(x), \quad \text{where} \quad n(x) = \frac{1}{e^{\beta x} - 1}. \tag{8.94}$$

Suppose the sum

$$S_1 = \frac{1}{\beta}\sum_n \frac{1}{i\omega_n - \omega_q}\frac{1}{i(\omega_m + \omega_n) + \omega_k}, \tag{8.95}$$

which can be written as

$$S_1 = \frac{1}{\beta}\sum_n \frac{1}{i\omega_m + \omega_q + \omega_k}\left[\frac{1}{i\omega_n - \omega_q} - \frac{1}{i\omega_n + i\omega_m + \omega_k}\right]. \tag{8.96}$$

Using (8.94) we write

$$S_1 = \frac{1}{i\omega_m + \omega_q + \omega_k}[n(\omega_q) - n(-i\omega_m - \omega_k)]. \tag{8.97}$$

Using $e^{2im\pi} = 1$ we get

$$S_1 = \frac{1}{i\omega_m + \omega_q + \omega_k}[1 + n(\omega_q) + n(\omega_k)]. \tag{8.98}$$

In the same way, we can show that if

$$S_2 = \frac{1}{\beta}\sum_n \frac{1}{(i\omega_n - \omega_q)} \frac{1}{[i(\omega_m + \omega_n) - \omega_k]}, \tag{8.99}$$

we find

$$S_2 = \frac{1}{i\omega_m + \omega_q - \omega_k}[n(\omega_q) - n(\omega_k)]. \tag{8.100}$$

8.9 Magnon–phonon interaction

As a simple example, I will study the one-dimensional ferromagnet coupled to phonons to show how we can go straight to the point. I will be mainly interested in the steps involved and not in the details of the calculations. At non-zero temperatures, the ground state's order is disrupted by thermally excited lattice vibrations, which are quantized as *phonons*. To show how the interaction arises, we start with the coupling exchange written as $J_{ij} = J(|x_{ij}|)$, where $x_{ij} = x_i - x_j$ is the distance between neighboring sites i and j. Then, we write $x_{ij} = x_{ij}^{(0)} + \delta x_i - \delta x_j$ where $\delta x_{i,j}$ are the instantaneous displacements of the respective ions. We can expand these displacements as linear combinations of phonons' creation and annihilation operators

$$\delta x_i = \frac{C}{\sqrt{N}}\sum_k \frac{1}{\sqrt{\Omega_k}}(c_k^\dagger - c_{-k})e^{-kx_i^{(0)}} \tag{8.101}$$

where C is a constant that depends on the ion mass and Ω_k is the phonon frequency. Next, we expand the exchange interactions in a Taylor series around the equilibrium distance $x_{ij}^{(0)}$

$$J(x_{ij}) \approx J + \frac{\partial J}{\partial x_{ij}} \Big| x_{ij}^{(0)}(\delta x_i - \delta x_j). \tag{8.102}$$

Substituting (8.102) into H, and considering only the quadratic term for the magnons we get

$$H = \sum_k \omega_k a_k^\dagger a_k + \sum_k \Omega_k(c_k^\dagger c_k + 1/2) + H_{mp}, \tag{8.103}$$

where

$$H_{mp} = \sum_{k,q} f(k, q) a^\dagger_{k-q} a_k (c^\dagger_q - c_{-q}), \tag{8.104}$$

is the magnon–phonon interaction. The noninteracting Green function for the magnon is

$$G^{(0)}(k, t) = -i\langle 0 | T a_k(t) a^\dagger_k(0) | 0 \rangle, \quad G^{(0)}(k, \omega) = \frac{1}{\omega - \omega_k + i\eta}. \tag{8.105}$$

For the phonon, we define an operator $\phi_k = c_k + c^\dagger_{-k}$ and write the Green function as

$$D^{(0)}(k, t) = -i\langle 0 | T \phi_k(t) \phi^\dagger_k(0) | 0 \rangle. \tag{8.106}$$

The Fourier transform is given by

$$D^{(0)}(k, \omega) = \frac{1}{\omega - \Omega_k + i\eta} - \frac{1}{\omega + \Omega_k - i\eta} = \frac{2\Omega_k}{\omega^2 - \Omega_k^2 + 2i\Omega_k \eta}. \tag{8.107}$$

Our goal is to calculate the Green function $D(k, \omega)$ for the interacting phonons. We start at zero temperature. Looking into equation (8.37), we see that the first contribution to $D^{(0)}(k, \omega)$ comes from the term depicted in figure 8.5, where we represent the magnon propagator by a solid line and the phonon propagator by a wavy line. Note that contractions between magnon and phonon operators vanish.

The self-energy expression corresponding to this diagram is

$$\Sigma(q, \omega) = \sum_k F(k, q) \int_{-\infty}^{\infty} d\omega' G^{(0)}(k - q, \omega - \omega') G^{(0)}(k, \omega'), \tag{8.108}$$

or, written explicitly

$$\Sigma(q, \omega) = \sum_k F(k, q) \int_{-\infty}^{\infty} d\omega' \frac{1}{\omega - \omega' - \omega_{k-q} + i\eta} \frac{1}{\omega' - \omega_k + i\eta}. \tag{8.109}$$

We will not be interested in the form of the function $F(q, \omega)$. Following the procedure presented in section (8.8), we get the expression for finite temperature

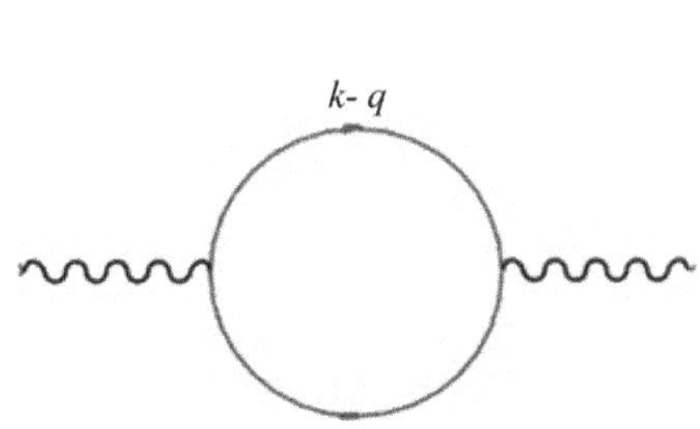

Figure 8.5. Phonon–magnon interaction.

$$\Sigma(q, \omega) = \sum_k F(k, q) \frac{1}{\omega + \omega_q - \omega_{k-q}} [n(\omega_q) - n(\omega_k)]. \tag{8.110}$$

As we can see, $\Sigma(q, \omega)$ vanishes at zero temperature. We can perform the sum in k numerically.

8.10 Exercise

8.1. Draw the next order diagram in figure 8.1. Write the corresponding term in the Green function expansion.

References

[1] Abrikosov A A, Gorkov L P and Dzyaloshinski I E 1963 *Methods of Quantum Field Theory in Statistical Physics* (New York: Dover Publications)

[2] Fetter A L and Walecka J D 2003 *Quantum Theory of Many-Particle Systems* (New York: Dover Publications)

[3] Mahan G D 1990 *Many-Particle Physics* (New York: Plenum)

[4] Lurie D 1968 *Particles and Fields* (New York: Interscience Publishers)

[5] Ziman J M 1969 *Elements of Advanced Quantum Theory* (Cambridge: Cambridge University Press)

[6] Altland A and Simons B 2006 *Condensed Matter Field Theory* (Cambridge: Cambridge University Press)

[7] Canali C M and Girvin S M 1992 Theory of Raman scattering in layered cuprate materials *Phys. Rev.* B **45** 7127–60

[8] Sentef M, Kolar M and Kampf A P 2007 Spin transport in Heisenberg antiferromagnet in two and three dimensions *Phys. Rev.* B **75** 214403

Antonio Sergio Teixeira Pires

Chapter 9

Topological magnon Hall effects

In this chapter, I will present a brief introduction to topological effects in magnetic systems. With the material introduced here, I hope the reader can start doing calculations in problems related to the subject.

9.1 Quantum Hall effect of electrons

Suppose there is a conducting surface in the xy-plane in which a current I is directed along the x-direction. This current produces an electric field E_x, and the current density is related to the field by $j_x = \sigma_0 E_x$. Then we apply a magnetic field oriented upward along the z-axis. The electrons will be deflected sideways in the negative y-direction. Suppose the surface is constrained in the y-direction. In that case, electrons deflected by the magnetic field will ultimately run into the edges of the surface, and they will produce an electric field E_y, which will point in the positive y-direction. This accumulation of electrons at the edges stops when E_y is sufficiently large to cancel the Lorentz force, and a current will occur only along the x-axis. E_y is referred to as the Hall field. The transverse conductivity defined by $j_x = \sigma_{xy} E_y$ is called the Hall conductivity [1]. In 1980, von Klitzing and collaborators [2] showed that σ_{xy} was quantized. A system must break the time-reversal symmetry to support the quantum Hall effect. That is done using an external magnetic field $\vec{B}$ or some other method. In the quantum Hall effect, electrons in the top edge move to the right. At the bottom, edge electrons flow to the left in the exact opposite direction. We say that the states are *chiral* (an object is called chiral if it is distinguishable from its mirror image). Chirality here is a consequence of time-reversal symmetry breaking.

In the *quantum spin Hall effect*, time-reversal symmetry is preserved due to the absence of external magnetic fields. Spin–orbit coupling generates an effective magnetic field that acts upward on the up-spin and the downward direction on the down-spin. As a result, electrons with an upward spin move (in a separate conducting channel) opposite to the spin-down electrons at the edge of the sample.

doi:10.1088/978-0-7503-3879-0ch9

However, there is no net flow of charge (only a net spin current) [3]. In the quantum spin Hall effect, in the top edge of the sample, spin-up electrons move in one direction (say, right), and spin-down electrons move in the opposite direction (left), and vice versa at the bottom edge. We say that the modes now are *helical*.

A topological insulator is a material that behaves as an insulator in the bulk but whose surface or edges contain conducting states. Hence, the main characteristic of topological insulators is that they can support chiral edge/surface states which are absent in conventional insulators. The bulk-boundary correspondence establishes a relation between a bulk property (for instance, gapped modes in the bulk) and what happens at its boundary (in this case, gapless modes that live on the edge). Small perturbations cannot destroy the edge modes, and we say that they are topologically protected. More details can be found in the literature (see, for instance [1], for a short and clear presentation of the subject).

9.2 Magnons in ferromagnets

A moving magnetic dipole moment $g\mu_B\hat{e}_z$ interacts with an external electric field and, analogous to the Aharonov–Bohm effect of charged particles in a magnetic field, acquires an Aharonov–Casher (AC) phase [4] given by

$$\theta_{ij} = (g\mu_B/\hbar c^2)\int_{x_i}^{x_j} d\vec{r}\cdot(\vec{E}\times\hat{e}_z). \tag{9.1}$$

In the following, I consider a two-dimensional ferromagnet in the xy-plane in the presence of an electric field $\vec{E}$ that couples to the magnetic dipole of the spins. The Hamiltonian is then written as [5]

$$H = -\sum_{<ij>} J_{ij}\left[\frac{1}{2}(S_i^+S_j^-e^{i\theta_{ij}} + S_i^-S_j^+e^{-i\theta_{ij}}) + S_i^zS_j^z\right], \tag{9.2}$$

where θ_{ij} is the AC phase, which the magnetic dipole moment, associated with the spin along the z-direction, acquires when it hops between neighboring sites.

We saw in section 3.3, equation (3.32), that the Hamiltonian of a ferromagnet with the conventional Dzyaloshinskii–Moriya interaction with the DM vector given by $\vec{D}_{ij} = D_{ij}\hat{e}_z$ could be written as an expression similar to equation (9.2). Therefore, the DM vector $\vec{D}$ acts as an electric field. As was said before, the DM interaction exists when the inversion symmetry of the lattice is broken at the middle point between the two spins. Hence, when the crystal structure has the inversion symmetry, an external electric field $\vec{E}$ can be used to induce an effect similar to the DM interaction. Namely, $\vec{D}_{ij} \propto \vec{E}\times\vec{e}_{ij}$, where $\vec{e}_{ij}$ is the unity vector connecting the two sites i and j [6, 7].

Using the Holstein–Primakoff transformation, we can write (9.2) as

$$H = -\sum_{<ij>} t_{ij}(e^{i\theta_{ij}}a_ia_j^\dagger + h.\ c.), \tag{9.3}$$

where $t_{ij} = J_{ij}S$ is the hopping amplitude.

The effective Hamiltonian for a magnetic moment in an external electric field is

$$H_m = \frac{1}{2m}\left(\vec{p} + \frac{g\mu_B}{c}\vec{A}(\vec{r})\right)^2, \tag{9.4}$$

where

$$\vec{A}(\vec{r}) = \frac{1}{c}\vec{E}(\vec{r}) \times \hat{e}_z, \tag{9.5}$$

m is an effective mass of the magnons and $\vec{p} = (p_x, p_y, 0)$ is the momentum operator of the magnon in the xy-plane. From the canonical equations of motion $\dot{\vec{r}} = \vec{v} = \partial H/\partial\vec{p}$ and $\dot{\vec{p}} = -\partial H/\partial\vec{r}$, one obtains the force acting on magnons due to the interaction with the electric field as [5, 8, 9]

$$\vec{F} = -g\mu_B\frac{\vec{v}}{c} \times (\vec{\nabla} \times \vec{A}(\vec{r})). \tag{9.6}$$

Let us take $\vec{E} = E_0(x, y, -2z)$ with E_0 a constant. Then, the magnons' equation of motion is formally identical to that of electrons in a constant magnetic field. Magnons are deflected into the $\hat{e}_y$ direction perpendicular to the transport direction $\hat{e}_x$. Since the mathematical structure of magnons in the presence of an AC phase is similar to that of electrons with an AB phase, we can use the topological formulation of the quantum Hall effect for electrons to study magnons. The existence of a spin Hall effect demands that the time-reversal symmetry should be broken. Here, it is broken by the electric field perpendicular to the xy-plane and by the finite magnetization.

Recently there has been considerable interest in the theoretical and experimental studies of topological magnon insulators (TMI) [10–12]. Like topological insulators in electronic systems, TMI are also characterized by non-trivial edge state modes when there is a gap in the bulk. In these systems, one has a transverse heat current mediated by magnons in response to a thermal gradient or a transverse spin current in response to a magnetic field gradient. In analogy with electronic systems, these currents are named thermal Hall and spin Hall currents, respectively. On the other side, the spin Nernst conductivity relates the spin current to a temperature gradient. The thermal magnon Hall effect was first predicted by Katsura *et al* [7] and verified experimentally by Onose *et al* [13] in the pyrochlore magnet $Lu_2V_2O_7$. Noninteracting topological magnons, being uncharged, can propagate for a long time without dissipation. This property makes them an excellent candidate to be used in spintronics. Because of the long-range coherence of spin waves, magnon Hall transport is more effective in device applications than electronic systems, and a new area of research, called magnonics, has appeared [14].

9.3 Transport in two-band models

In this section, I study a generic ferromagnetic two-band system in two dimensions described by the Hamiltonian

$$H = \sum_k \psi_k^{\dagger} M_k \psi_k, \tag{9.7}$$

where $\psi_k^{\dagger} = (a_k^{\dagger}, b_k^{\dagger})$ and the matrix M_k is given by

$$M_k = h_0(k)I + h_x(k)\sigma_x + h_y(k)\sigma_y + h_z(k)\sigma_z. \tag{9.8}$$

In a Hamiltonian with time-reversal symmetry, we should have $h_y = 0$, since the time-reversal operation changes i into $-i$. The matrix M_k can be written as

$$M_k = h_0(k)I + \begin{pmatrix} h_z & h_x - ih_y \\ h_x + ih_y & -h_z \end{pmatrix}. \tag{9.9}$$

The eigenvalues of M_k are given by

$$\omega_{\pm}(k) = h_0(k) \pm h(k) \tag{9.10}$$

where

$$\vec{h}(k) = (h_x(k), h_y(k), h_z(k)), \;\; h(k) = \sqrt{h_x^2(k) + h_y^2(k) + h_z^2(k)}. \tag{9.11}$$

The eigenvectors of M_k are

$$|k, +\rangle = \begin{pmatrix} p_1 e^{-i\varphi} \\ p_2 \end{pmatrix}, \;\; |k, -\rangle = \begin{pmatrix} p_2 e^{-i\varphi} \\ -p_1 \end{pmatrix}, \tag{9.12}$$

where

$$p_1 = \sqrt{\frac{h + h_z}{2h}}, \;\; p_2 = \sqrt{\frac{h - h_z}{2h}}, \;\; \cos\varphi = \frac{h_x}{\sqrt{h_x^2 + h_y^2}}, \;\; \sin\varphi = \frac{h_y}{\sqrt{h_x^2 + h_y^2}}. \tag{9.13}$$

In analogy with charge transport, the spin-current operator is given by [17]

$$J_\alpha(k) = \frac{\partial M_k}{\partial k_\alpha} = \frac{\partial h_0}{\partial k_\alpha} + \sum_i \frac{\partial h_i}{\partial k_\alpha}\sigma_i, \tag{9.14}$$

where $\alpha = x, y$, $i = x, y, z$.

Using the Kubo formula equation (7.93), we arrive at the following expression for the transverse static conductivity $\sigma_{xy}(0)$, called the spin Hall conductivity

$$\sigma_{xy} = \frac{i}{A}\sum_k \sum_{m \neq n} \frac{(n_k^n - n_k^m)\,\mathrm{Im}\langle k, n|J_x(k)|k, m\rangle\langle k, m|J_y(k)|k, n\rangle}{[E_n(k) - E_m(k)][E_n(k) - E_m(k) + i\eta]}, \tag{9.15}$$

where n_k^n is the Bose factor and A is the area of the system, and for the model studied here, we have $E_{\pm} = \omega_{\pm}$. As we saw in chapter 7, σ_{xy} relates the spin current in the y-direction to a field gradient in the x-direction. As in chapter 7, The positive infinitesimal η has been introduced to make time integrals well defined. We limit our study to the case that the two levels do not cross in the whole Brillouin zone such that we can take $\eta \to 0^+$ before we integrate in $\vec{k}$.

For $n \neq m$ we can write

$$\langle k, n|J_\alpha(k)|k, m\rangle = \sum_i \frac{\partial h_i}{\partial k_\alpha}\langle k, n|\sigma_i|k, m\rangle. \tag{9.16}$$

From (9.12) we get

$$\begin{aligned}
&\langle k, +|\sigma_z|k, -\rangle = \langle k, -|\,\sigma_z|k, +\rangle = 2p_1p_2,\\
&\langle k, +|\sigma_x|k, -\rangle = -p_1^2e^{i\varphi} + p_2^2e^{-i\varphi},\\
&\langle k, -|\sigma_y|k, +\rangle = p_2^2e^{i\varphi} - p_1^2e^{-i\varphi}\\
&\langle k, +|\sigma_y\,|k, -\rangle = i(p_1^2e^{i\varphi} + p_2^2e^{-i\varphi}),\ \langle k, -|\sigma_y|k, +\rangle = -i(p_2^2e^{i\varphi} + p_1^2e^{-i\varphi}),
\end{aligned} \tag{9.17}$$

which leads to

$$\mathrm{Im}\langle k, n|\sigma_i|k - n\rangle\langle k, -n|\sigma_j|k, n\rangle = n\varepsilon_{ijk}\frac{h_k}{h}. \tag{9.18}$$

The final result for the spin Hall conductivity is

$$\sigma_{xy} = -\frac{1}{2A}\sum_k \varepsilon_{ijk}\frac{(n_k^+ - n_k^-)}{h^3}\left(\frac{\partial h_i}{\partial k_x}\right)\left(\frac{\partial h_j}{\partial k_y}\right)h_k, \tag{9.19}$$

which can be written as

$$\sigma_{xy} = -\frac{1}{2A}\int_{BZ} dk_xdk_y(n_k^+ - n_k^-)\Omega(\vec{k}), \tag{9.20}$$

where

$$\Omega(\vec{k}) = \frac{(\partial_{k_x}\vec{h} \times \partial_{k_y}\vec{h}).\,\vec{h}}{2h^3}. \tag{9.21}$$

$\Omega(\vec{k})$ is the Berry curvature and the integral is over the whole Brillouin zone. I will define all terms so that the reader can follow all of the material presented here, even if they have not studied topology. We associate a Berry curvature, with an opposite sign, to each band. Then we put an index in Ω and write $\Omega_n(\vec{k})$, with $n = \pm$. With this notation, we can write σ_{xy} as

$$\sigma_{xy} = \sum_{\lambda=\pm}\frac{1}{(2\pi)^2}\int_{BZ} dk_xdk_yn_k^\lambda\Omega_\lambda(\vec{k}). \tag{9.22}$$

Where we consider now $\Omega_\pm(\vec{k})$ as the Berry curvature of the upper and lower magnons' bands. Performing the calculation, we find

$$\Omega_+(\vec{k}) = \frac{1}{h^3}\left\{h_x\left(\frac{\partial h_y}{dk_x}\frac{\partial h_z}{\partial k_y} - \frac{\partial h_z}{\partial k_x}\frac{\partial h_y}{\partial k_y}\right) + h_y\left(\frac{\partial h_z}{\partial k_x}\frac{\partial h_x}{\partial k_y} - \frac{\partial h_x}{\partial k_x}\frac{\partial h_z}{\partial k_y}\right)\right.$$
$$\left. + h_z\left(\frac{\partial h_x}{\partial k_x}\frac{\partial h_y}{\partial k_y} - \frac{\partial h_y}{\partial k_x}\frac{\partial h_x}{\partial k_y}\right)\right\}$$

and $\Omega_-(\vec{k}) = -\Omega_+(\vec{k})$. We see that the condition $\sigma_{xy} \neq 0$ demands $h_x \neq 0$, $h_y \neq 0$, $h_z \neq 0$, and a finite gap (in equation (9.15) we should have $E_n(k) \neq E_m(k)$ for all $\vec{k}$). A direct consequence of non-zero Berry curvature in a magnonic system is a spin Hall current perpendicular to an applied field gradient.

The integral of the Berry curvature is the Chern number

$$C^{\pm} = \frac{1}{2\pi}\int_{BZ} \Omega_{\pm}(\vec{k})dk_x\, dk_y. \tag{9.23}$$

This expression is the well-known result [3] that the Chern number of a two-band model is equivalent to the winding number of the mapping from a two-dimensional Brillouin zone, which is a 2D torus to the 2D unit sphere. If h_x, h_y, and h_z are independent of each other, and each of them can take both positive and negative values when the other two components are zero, then $\vec{h}$ will sweep out the whole unit sphere, which encloses the singular point inside. Hence the winding number is non-trivial. But if $\vec{h}$ winds the sphere twice in opposite directions and cancels out or comes back to a given point as $\vec{k}$ is varied over all its values without sweeping the entire sphere, the Chern number is zero [15]. For $C \neq 0$ a monopole, singularity must exist, and this condition requires that $h_z(\vec{k})$ goes through zero at some point $\vec{k}$ and change sign as that point is crossed. If $h_z(\vec{k}) = 0$ for all $\vec{k}$, the sphere turns to a circle, and no singularity can be defined.

Notice that we need at least two bands to have a non-trivial topology, and therefore, a lattice with at least two atoms per cell. As a consequence, single-band models have only trivial Chern numbers. We remark that, in contrast to the case of electrons, there are no filled bands in the bosonic system. Therefore, each magnon band contributes to the current, and non-zero Chern numbers indicate only the existence of topologically protected chiral magnon edge states. Another difference with fermions is that boson numbers are not conserved and vanish at zero temperature.

A ferromagnet (or antiferromagnet) with a magnon spectrum with non-zero Chern number is called a *topological magnon insulator*, although the term 'insulator' has no physical meaning for magnons.

Now we can calculate the Berry curvature for an N band system. Using equations (9.15) and (9.14) we get

$$\Omega_n(\vec{k}) = i\sum_{m\neq n} \frac{\langle k, m|\nabla_k M_k|k, n\rangle\langle k, n|\nabla_k M_k|k, m\rangle}{[E_m(\vec{k}) - E_n(\vec{k})]^2}. \tag{9.24}$$

This expression is sometimes used in the literature as the definition of the Berry curvature. We can also insert $\sum_{m'}|k, m'\rangle\langle k, m'| = I$ into (9.24) and obtain the term

$$\langle k, n \,|\, \frac{\partial}{k_\alpha} \,|k, m'\rangle\langle k, m'|\; M_k \,|k, m\rangle. \;\; (\alpha = x, y). \tag{9.25}$$

Renaming the eigenvectors as u_{nk}, it is easy to show that we can write

$$\Omega_n(\vec{k}) = -i\left(\left\langle \frac{\partial u_{nk}}{\partial k_\alpha} \middle| \frac{\partial u_{nk}}{\partial k_\beta} \right\rangle - \left\langle \frac{\partial u_{nk}}{\partial k_\beta} \middle| \frac{\partial u_{nk}}{\partial k_\alpha} \right\rangle\right). \;\; (\alpha, \beta = x, y) \tag{9.26}$$

where $n = 1,2,\ldots,N$. Both equations (9.24) and (9.26) are general expressions for the Berry curvature independent of the model used. It can be shown that the sum of Chern numbers over all bands is zero [16]. The Berry curvature and Chern numbers have the following properties:

(i) For systems with time-reversal symmetries $\Omega(\vec{k})$ is an odd function of $\vec{k}$. So, the Chern number for a time-reversal invariant system must be zero.
(ii) For a system invariant under space inversion, the Berry curvature is an even function of $\vec{k}$.
(iii) If a system has both time-reversal and space inversion symmetry, the Berry curvature must be zero at every $\vec{k}$ point.

9.4 Thermal Hall conductivity

A longitudinal temperature gradient can generate a transverse spin current (spin Nernst effect) or a transverse heat current. Typically, one associates the observation of a transverse current with the term Hall effect. The linear-response theory for the energy current can be developed following the procedure presented in chapter 7. The energy density is coupled to a pseudo gravitational potential as an effective way to derive the thermal response function. As we saw in chapter 7, the source term for the spin current, $-\sum_i h_i S_i^z$, does not modify the spin continuity equation since $[S_i^z, S_j^z] = 0$. That is not the case here, and the calculations are more complex. I will present only the expressions for the transport coefficients. For details of the calculations, see Matsumoto *et al* [11] and Han *et al* [17]. The spin Nernst coefficient and thermal Hall conductivity are given, respectively, by (setting $k_B = \hbar = 1$)

$$\alpha_{xy} = \sum_\lambda \frac{1}{(2\pi)^2} \int_{BZ} dk_x dk_y c_1(n_k^\lambda)\Omega_\lambda(\vec{k}), \tag{9.27}$$

$$\kappa^{xy} = -\sum_\lambda \frac{T}{(2\pi)^2} \int_{BZ} dk_x dk_y c_2(n_k^\lambda)\Omega_\lambda(\vec{k}), \tag{9.28}$$

where

$$c_1(x) = (1 + x)\ln(1 + x) - x\ln x, \tag{9.29}$$

$$c_2(x) = (1 + x)\left[\ln\left(\frac{1 + x}{x}\right)\right]^2 - (\ln x)^2 - 2Li_2(-x). \tag{9.30}$$

Li_2 is the dilogarithm function given by

$$Li_2(x) = \sum_{n=1}^{\infty}\frac{x^n}{n^2}, \quad \text{for } |x| \leqslant 1, \tag{9.31}$$

$$Li_2(-x) = \frac{\pi^2}{16} - \frac{1}{2}\ln^2 x + \sum_{n=1}^{\infty}\frac{(-1)^{n-1}}{n^2x^n}, \text{ for } x > 1. \tag{9.32}$$

The function $c_2(x)$ increases monotonically with x, it is zero at $x = 0$ and tends to $\pi^3/3$ as $x \to \infty$. As $\Omega_\lambda(\vec{k})$ is weighted by $c_2(n_k^\lambda)$, there is a non-vanishing thermal Hall conductivity at finite temperature even if the two bands have zero Chern numbers. In the spin-wave formalism, the temperature dependence of k^{xy} rests only on $c_2(n_k^\lambda)$, and k^{xy} tends to a constant value at high temperatures where both bands are equally occupied. As $T \to 0$, both bands are nearly empty and $k^{xy} \to 0$.

The conductivities coefficients are expected to be much smaller for a system with $C = 0$, than for one with non-zero Chern numbers.

One example of a two-band model with non-trivial topology is the two-dimensional honeycomb ferromagnetic lattice studied in chapter 4, with a Dzyaloshinskii–Moriya (DM) term, described by the Hamiltonian [18, 19]

$$H = -J\sum_{<ij>}\vec{S}_i.\,\vec{S}_j + D\sum_{<<ij>>}\nu_{ij}\hat{z}.\,(\vec{S}_i \times \vec{S}_j). \tag{9.33}$$

The second term is the DM interaction between next nearest neighbors (NNN), where the constants $\nu_{ij} = -\nu_{ji} = \pm 1$ characterize the dependence of the interaction on the relative position of the two next nearest-spins. An out-of-plane DMI ($\vec{D}_{ij} = \nu_{ij}D\hat{z}$) is allowed due to the lack of an inversion center at the mid-point between NNN bonds. The structure factor γ_k is complex (as was shown in chapter 4) and is given by:

$$\gamma_k = e^{ik_x/2}[2\cos(\sqrt{3}\,k_x/2) + e^{-3ik_y/2}]. \tag{9.34}$$

Using the Holstein–Primakoff representation, we obtain the noninteracting Hamiltonian (9.9) with $h_x + ih_y = \gamma_k$. The h_z term comes from the DM interaction and is given by

$$h_z = 2DS\sin\left(\frac{\sqrt{3}}{2}k_x\right)\left[\cos\left(\frac{\sqrt{3}k_x}{2}\right) - \cos\left(\frac{3k_y}{2}\right)\right]. \tag{9.35}$$

The dispersion relation in the absence of the DM term was given in equation (4.14). The upper and the lower band meet at the points $\vec{K} = (4\pi/3,0)$ and $\vec{K}' = (2\pi/3,2\pi/\sqrt{3})$ forming linearly-dispersed bands. The DM term breaks the lattice's inversion symmetry and opens an energy gap at these points, making the band structure topologically non-trivial. The DM interaction also leads to a phase term in the Hamiltonian (see equation 3.32), so that this term plays the role of the external electric field in equation (9.2). Kim [18] and Owerre [19] calculated the transport constants for the honeycomb lattice (and also have shown that $C^{\pm} = \pm 1$).

A drawback of the Holstein–Primakoff approach is that it works poorly in the weakly ordered regime and fails for the paramagnetic phase. A paramagnetic state with restored time-reversal symmetry should have zero Hall conductivities anyway, but by applying an external magnetic field, one can restore the Hall transport.

9.5 Three-band model

Now we consider the case when the matrix M is given by

$$M = R_0 I + \vec{R}.\,\vec{L}, \tag{9.36}$$

where $\vec{R} = (R_x, R_y, R_z)$, $\vec{L} = (L_x, L_y, L_z)$, and L_x, L_y, L_z are the spin-1 representation of the SU(2) generators given by

$$L_x = \begin{pmatrix} 0 & 1 & 0 \\ 1 & 0 & 0 \\ 0 & 0 & 0 \end{pmatrix},\ L_y = \begin{pmatrix} 0 & 0 & 1 \\ 0 & 0 & 0 \\ 1 & 0 & 0 \end{pmatrix},\ L_z = \begin{pmatrix} 0 & 0 & 0 \\ 0 & 0 & -i \\ 0 & i & 0 \end{pmatrix}, \tag{9.37}$$

and I is the 3×3 unit matrix. The matrix M is then written as

$$M = \begin{pmatrix} R_0 & R_x & R_y \\ R_x & R_0 & -iR_z \\ R_y & iR_z & R_0 \end{pmatrix}. \tag{9.38}$$

To obtain the eigenvalues, we calculate $\det(M - \lambda I) = 0$ and obtain

$$(R_0 - \lambda)^3 - (R_x^2 + R_y^2 + R_z^2)(R_0 - \lambda) = 0, \tag{9.39}$$

which leads to

$$E_1 = R_0 + R,\ E_2 = R_0,\ E_3 = E_0 - R, \tag{9.40}$$

with $R = \sqrt{R_x^2 + R_y^2 + R_z^2}$.

E_1, and E_3 refer to the lower and upper dispersive band, respectively, and E_2 is a dispersionless flat band. The corresponding eigenvectors are found to be

$$
\begin{aligned}
|\psi_2\rangle &= \frac{1}{R}\begin{pmatrix} iR_z \\ R_y \\ -R_x \end{pmatrix}, \\
|\psi_1\rangle &= \frac{1}{R\sqrt{2(R^2 - R_x^2}}\begin{pmatrix} -RR_y + iR_xR_z \\ R_xR_y - iRR_z \\ R^2 - R_x^2 \end{pmatrix}, \\
|\psi_3\rangle &= \frac{1}{R\sqrt{2(R^2 - R_x^2}}\begin{pmatrix} RR_y + iR_xR_z \\ R_xR_y + iRR_z \\ R^2 - R_x^2 \end{pmatrix}.
\end{aligned}
\tag{9.41}
$$

We can verify that

$$
\langle\psi_i||\psi_i\rangle = 1 \text{ and } \langle\psi_i|\big|\psi_j\big\rangle = 0 \quad \text{for } i \neq j.
$$

There is a freedom in choosing the eigenvectors related by U(1) gauge transformations. However, the gauge does not affect the Berry curvature. The Berry curvature for the three bands is given by

$$
\Omega_2(\vec{k}) = 0, \quad \Omega_{1,3}(\vec{k}) = \mp\frac{\vec{R}}{R^3}\cdot\left(\frac{\partial\vec{R}}{\partial k_x} \times \frac{\partial\vec{R}}{\partial k_y}\right). \tag{9.42}
$$

Note that there is a factor of 2 difference compared to the two-band model (equation 9.21). The above result can be generalized to any spin-$n/2$ representation of SU(2) for integer n [15]. The Kagome [12, 20] and the Lieb lattices [21] have three magnon bands, and the Shastry–Sutherland lattice [14] has four distinct magnons bands.

The Lieb lattice is shown in figure 9.1. It has a unit cell with three nonequivalent sites A, B and C. Therefore, we have to use three boson operators (a_i, b_i, c_i). Site A has four nearest neighbors (NN), namely two B and two C sites. B and C sites have only two NN A sites. The Hamiltonian of our interest is given by

$$
H = -J\sum_{<m,n>}\vec{S}_m\cdot\vec{S}_n - D\sum_{<<m,n>>}(S_m^xS_n^y - S_m^yS_n^x), \tag{9.43}
$$

where J is the nearest-neighbor exchange coupling between sites AB and AC, and D is the Dzyaloshinskii–Moriya interaction (DMI) between next nearest-neighbor (NNN) sites CB along the diagonal. In a lattice with inversion symmetry, the conventional DMI is absent. However, as was said before, this term can be induced by an external electric field E. As it is usually done in the literature, we consider the case where the NNN magnons hopping is purely imaginary, which leads to non-trivial topological effects. Using the Holstein–Primakoff representation, we write the Hamiltonian as

$$
\begin{aligned}
H_1 = &-JS\sum_{i,j}[a_i^\dagger b_j + b_i^\dagger a_j + a_i^\dagger c_j + c_i^\dagger a_j - (a_i^\dagger a_i + b_i^\dagger b_i + c_i^\dagger c_i) \\
&-(a_j^\dagger a_j + b_j^\dagger b_j + c_j^\dagger c_j)] - iDS\sum_{ij}(a_i^\dagger c_j - c_i^\dagger a_j).
\end{aligned}
\tag{9.44}
$$

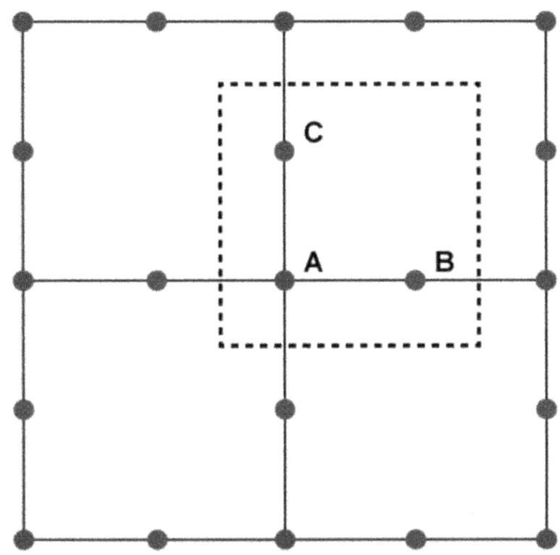

Figure 9.1. The Lieb lattice. Each cell contains three different sites A, B and C. $a = 1$ is the lattice parameter.

The topological properties of the Lieb lattice were studied by Cao *et al* [21].

The Lieb lattice stands out as the prototypical model for exploring flat band phenomena in optical lattices. A system with flat bands displays remarkable, strongly interacting phases of matter [22].

9.6 Calculation of the edge modes

As was said before, under the bulk-boundary correspondence, if there is a gap in the bulk of a topological insulator (when we treat an infinite lattice), there are gapless modes in the border of a finite sample of the system. To illustrate the procedure of how to calculate the edge modes, I will consider a simple Hamiltonian in two dimensions, describing a ribbon with a finite number of chains in the y-direction, but retaining periodic boundary conditions in the x-direction [23]. The Hamiltonian of my interest is given by

$$H = \sum_{k_x,k_y} (A + B\sin k_y + C\cos k_y)a_k^\dagger a_k, \tag{9.45}$$

where A, B and C may be functions of k_x. Of course, this Hamiltonian is topologically trivial, and the aim here is only to show how to perform the calculation. We write

$$a_k = \frac{1}{L}\sum_n e^{ik_y n} a_n(k_x), \tag{9.46}$$

where n is the coordinate in the y-direction: $n = 0, 1, 2, \ldots, L$.

To proceed, we write H as

$$H = \frac{1}{L^2}\sum_{k_x,k_y}\left\{\left[A + B\frac{(e^{ik_y} - e^{-ik_y})}{2i} + C\frac{(e^{ik_y} + e^{-ik_y})}{2}\right]\right.$$
$$\left.\sum_{n,m} e^{-ik_y n} a_n^\dagger(k_x) e^{ik_y m} a_m(k_x)\right\} \tag{9.47}$$

or

$$H = \frac{1}{L^2}\sum_{k_x,k_y}\sum_{n,m}\left[A + \frac{B(e^{i(1-n+m)k_y} - e^{i(-1-n+m)k_y})}{2i} + \frac{C(e^{i(1-n+m)k_y} + e^{i(-1-n+m)k_y}}{2}\right] \tag{9.48}$$
$$a_n^\dagger(k_x)a_m(k_x).$$

Using the relation

$$\frac{1}{L}\sum_{k_y} e^{ik_y(m-n\pm1)} = \delta_{m-n,\,\pm1}, \tag{9.49}$$

and Fourier transforming back to real space in only the y-direction, we get

$$H = \frac{1}{L}\sum_{k_x}\sum_{n}[Aa_n^\dagger(k_x)a_n(k_x) + \tau a_n^\dagger(k_x)a_{n-1}(k_x) + \tau^* a_n^\dagger(k_x)a_{n+1}(k_x)], \tag{9.50}$$

where

$$\tau = \frac{C}{2} - i\frac{B}{2}. \tag{9.51}$$

For example, for $L = 4$ we have to diagonalize the matrix

$$H_{\text{ribbon}} = \begin{pmatrix} A & \tau^* & 0 & 0 \\ \tau & A & \tau^* & 0 \\ 0 & \tau & A & \tau^* \\ 0 & 0 & \tau & A \end{pmatrix}. \tag{9.52}$$

In general, we perform the calculation for a finite number of lines diagonalizing a matrix $L \times L$, numerically. In the case of an N band model, A and τ in equation (9.52) are matrices. The diagonalization of H_{ribbon} gives all the modes of the ribbon (bulk and edge modes).

9.7 Antiferromagnets

As we saw before, the magnon spin in a ferromagnet is fixed by its chirality, which can only be right-handed with respect to the magnetization. By contrast, in a collinear antiferromagnet, the symmetry of the model admits two degenerate magnon modes with opposite chirality, and hence opposite spin. The Hamiltonian of an antiferromagnet was studied in chapter 2. As we saw, we have two decoupled magnon modes with up and down spins. Therefore, the dynamics of noninteracting magnons in the AF can be described as the combination of two independent copies of the dynamics of magnons in a ferromagnet for each mode $\sigma = \pm 1$. For a magnon with a σ spin, a magnetic field gradient $\partial_x B$ along with the x-axis works as a driving force $\vec{F} = F_\sigma \hat{e}_x$ with $F_\sigma = \sigma g \mu_B \partial_x B$. Since the directions of the force are opposite for the two magnon modes $\sigma = \pm 1$, the magnetic field gradient generates helical magnon transport in the bulk. That is, magnons with opposite magnetic moments flow in the opposite x-direction. However, all magnons subjected to a thermal gradient flow in the same x-direction [8].

Up to now, there is nothing new. Now we study the effect of an electric field $\vec{E}$ that couples to the magnetic dipole moment $\sigma g \mu_B \hat{e}_z$ in a two-dimensional antiferromagnet in the xy-plane. As we saw in section 9.2, the magnon experiences a magnetic force in the magnon's rest frame in the presence of an electric field. The effective Hamiltonian is now given by $H_m = \sum_{\sigma=\pm 1} H_\sigma$, with

$$H_\sigma = \frac{1}{2m}\left(\vec{p} + \frac{\sigma g_{\mu_B}}{c^2}\vec{E}(\vec{r}) \times \hat{e}_z\right)^2 - \sigma g \mu_B B. \tag{9.53}$$

If the electric field is of the form $\vec{E}(\vec{r}) = \varepsilon(-x/2, -y/2, 0)$ with ε a constant field gradient, the field gradient ε plays the role of the perpendicular magnetic field in the two-dimensional electron gas (figure 9.2) [8]. The force acting in the magnon dipole moment is now given by

$$\vec{F} = \sigma g \mu_B\left(\nabla B - \vec{v} \times \frac{\varepsilon \hat{e}_z}{c^2}\right). \tag{9.54}$$

The motion can be treated as the superposition of a circular (cyclotron) motion with a velocity $\vec{v}_c$ around a point, and the drift of this point with velocity $\vec{v}_d$. If $\dot{\vec{v}}_d = 0$, equation (9.54) gives $\vec{v}_d \times \hat{e}_z = (c^2/\varepsilon)\nabla B$. If the magnetic field gradient is applied along the x-axis $\partial_x B \neq 0$ and at the same time, $\partial_y B = 0$, we get $\vec{v}_d = (0, c^2 \partial_x B/\varepsilon, 0)$, which is perpendicular to the applied magnetic field gradient and independent of σ (the σ dependence cancels out as $\sigma^2 = 1$). Hence, in the presence of both magnetic and electrical field gradients, each magnon ($\sigma = \pm 1$) performs the drift motion in the same direction. The drift velocity vanishes in the absence of the magnetic field

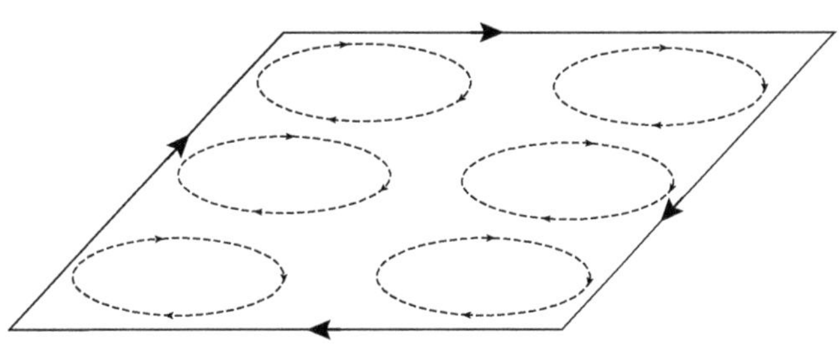

Figure 9.2. Electrically induced AC effect. Up magnons have a cyclotron orbit. The spin current in the bulk cancels out, but we have a chiral edge current. For down magnons, the direction of the current is the opposite.

gradient, while each magnon ($\sigma = \pm 1$) still shows the cyclotron motion in opposite directions due to the electric field gradient.

At low temperatures, only the lowest energy mode becomes relevant. So the cyclotron motion of up magnons along one direction leads to a chiral edge state with Chern number $C_\uparrow = +1$. The direction of cyclotron motion of the down magnons is opposite. They also propagate along the edge with Chern number $C_\downarrow = -1$ (note that up and down magnons propagate along the edge of the sample but in opposite directions). Then we have a bosonic analog of the quantum spin Hall effect for electrons. Note that we have $C_\uparrow + C_\downarrow = 0$.

For the benefit of the reader, I will review some topics that were presented in chapter 4. We start with a general Hamiltonian for an antiferromagnet on a lattice with two inequivalent sites

$$H = \frac{1}{2}\sum_k \psi_k^\dagger M_k \psi_k, \tag{9.55}$$

with $\psi_k^\dagger$ written as

$$\psi_k^\dagger = (a_k^\dagger, b_k^\dagger, a_{-k}, b_{-k}). \tag{9.56}$$

The Hamiltonian is diagonalized by the transformation

$$\phi_k = T_k \psi_k. \tag{9.57}$$

The eigenvalues and eigenvectors can be obtained diagonalizing the matrix

$$K_k = \eta M_k, \tag{9.58}$$

where $\eta = \mathrm{diag}(1,1,-1,-1)$. Then T_k is written as

$$T_k = [V(\varepsilon_{1k}), V(\varepsilon_{2k}), V(-\varepsilon_{1k}), V(-\varepsilon_{2k})], \tag{9.59}$$

where $V(\varepsilon_{ik})$ are the eigenvectors and ε_{ik} the eigenvalues. The Berry curvature is given by

$$\Omega_n(\vec{k}) = i \in_{xy} \left[\eta \frac{\partial T_k^\dagger}{\partial k_x} \eta \frac{\partial T_k}{\partial k_y}\right]_{nn}, \ n = 1,2, \tag{9.60}$$

where here $\in_{xy}$ is the antisymmetric tensor. We have $\Omega_3(\vec{k}) = -\Omega_1(\vec{k})$, and $\Omega_4(\vec{k}) = -\Omega_2(\vec{k})$. The thermal Hall conductivity is given by

$$\kappa_{xy} = -\frac{\kappa_B^2 T}{\hbar V} \sum_{n=1,2} \sum_k \left\{ c_2[n(\varepsilon_{nk})] - \frac{\pi^2}{3} \right\} \Omega_n(\vec{k}), \tag{9.61}$$

where the sum on n ran only over the positive energies and $c_2(\varepsilon_{nk})$ was defined in equation (9.30). The above results assume that $\psi_k^\dagger$ has been chosen as in (9.56). The sum of the Chern numbers over all particle bands is zero and thus

$$\sum_k \sum_n \left\{ -\frac{\pi^2}{3} \right\} \Omega_n(\vec{k}) = 0. \tag{9.62}$$

For this reason, we can neglect the term $-\pi^2/3$ in equation (9.61). The derivation of equation (9.61) with the Berry curvature given by equation (9.60), requires that M_k has been chosen to satisfy the particle-hole symmetry transformation:

$$M_k = \rho_1 M_{-k}^T \rho_1, \ \rho_1 = \begin{pmatrix} 0 & I_{N\times N} \\ I_{N\times N} & 0 \end{pmatrix}. \tag{9.63}$$

However, it may happen that using $\psi_k^\dagger = (a_k^\dagger, b_{-k}, a_{-k}, b_k^\dagger)$, as was the case for the honeycomb and checkerboard lattice studied in chapter 4, the Hamiltonian can be written in a block diagonal form as

$$M_k = \begin{pmatrix} M_{kI} & 0 \\ 0 & M_{kII} \end{pmatrix}, \tag{9.64}$$

where

$$M_{kI} = \begin{pmatrix} g_k & h_k \\ h_k^* & f_k \end{pmatrix}, \ M_{kI} = M_{kII}^*. \tag{9.65}$$

Defining

$$\Delta_k = \frac{g_k - f_k}{2}, \ R_k = \frac{g_k + f_k}{2}, \tag{9.66}$$

we get for the eigenvalues

$$\begin{aligned} \varepsilon_{1k} &= \Delta_k + \omega_k, \ \varepsilon_{3k} = \Delta_k - \omega_k, \\ \varepsilon_{2k} &= -\Delta_k + \omega_k, \ \varepsilon_{4k} = -\Delta_k - \omega_k, \end{aligned} \tag{9.67}$$

where $\omega_k = \sqrt{R_k^2 - | h_k |^2}$.

Defining

$$u_k = \cosh(\theta_k/2), \ v_k = \sinh(\theta_k/2), \ \tan \phi_k = \mathrm{Im} h_k / \mathrm{Re}\, h_k \tag{9.68}$$

where

$$\cosh \theta_k = \frac{R_k}{\omega_k}, \ \sinh \theta_k = \frac{|h_k|}{\omega_k}, \tag{9.69}$$

we find that the eigenvalues are given by [24]

$$T_{1k} = \begin{pmatrix} e^{i\phi_k} u_k \\ v_k \\ 0 \\ 0 \end{pmatrix}, \ T_{3k} = \begin{pmatrix} v_k \\ e^{-i\phi_k} u_k \\ 0 \\ 0 \end{pmatrix}, \ T_{2k} = \begin{pmatrix} 0 \\ 0 \\ e^{-i\phi_k} u_k \\ v_k \end{pmatrix}, \ T_{4k} = \begin{pmatrix} 0 \\ 0 \\ v_k \\ e^{i\phi_k} u_k \end{pmatrix}. \tag{9.70}$$

The antiferromagnetic honeycomb lattice with a Moriya–Dzyaloshinskii term was studied by Cheng *et al* [24], Zyuzin *et al* [25] and Owerre [26]. We have for this lattice

$$h_k = e^{ik_x/2}[2\cos(\sqrt{3}k_x/2) + e^{-3ik_y/2}], \tag{9.71}$$

and

$$g_k = 3JS + 2DSm_k,\ f_k = 3JS - 2DSm_k, \tag{9.72}$$

where

$$m_k = \sin\left(\frac{\sqrt{3}}{2}k_x\right)\left[\cos\left(\frac{\sqrt{3}k_x}{2}\right) - \cos\left(\frac{3k_y}{2}\right)\right]. \tag{9.73}$$

Note that m_k is an odd function of $\vec{k}$. The energy spectrum of the two magnon branches are given by (as mentioned before, we keep only the positive branch)

$$\varepsilon_{\alpha k} = S\left[J\sqrt{9 - |h_k|^2} + Dm_k\right],\ \varepsilon_{\beta k} = S\left[J\sqrt{9 - |h_k|^2} - Dm_k\right]. \tag{9.74}$$

The D term breaks the degeneracy. Using $m_{-k} = -m_k$ we see that the coefficient κ_{xy} vanishes.

In a honeycomb antiferromagnet in the presence of the DM interaction, both up-spin and down-spin magnons exhibit the magnon thermal effect under an applied in-plane longitudinal temperature gradient. Still, magnons with opposite spins flow in opposite transverse directions. The thermal Hall current thereby cancels out, while a non-zero spin current emerges in the transverse direction realizing a magnon Nernst effect (figure 9.3) [27]. The antiferromagnetic checkerboard lattice was studied by Pires [28].

9.8 Skyrmions

Position-dependent magnetization textures can be topologically non-trivial, and a variety of magnetic textures have been observed. Skyrmions are noncollinear spatially inhomogeneous spin textures in which the spin quantization axis changes continuously over length scales that vary from a few nm to a few μm. There are different definitions for this magnetic quasi-particle. A common definition is to define a skyrmion as being any spin structure in which the center magnetization is in the opposite direction to its boundary and which can be mapped to the sphere [29]. That is, the spin in the periphery point in the opposite direction to the spins in the center. Magnetic skyrmions exhibit unique real space topological characteristics and are considered as one of the most promising information carriers in spintronic devices.

A classification of spin structures can be based on their topology. Two-dimensional textures described by a local magnetization $\vec{m}$ can be classified by the skyrmion winding number, or topological charge [30]

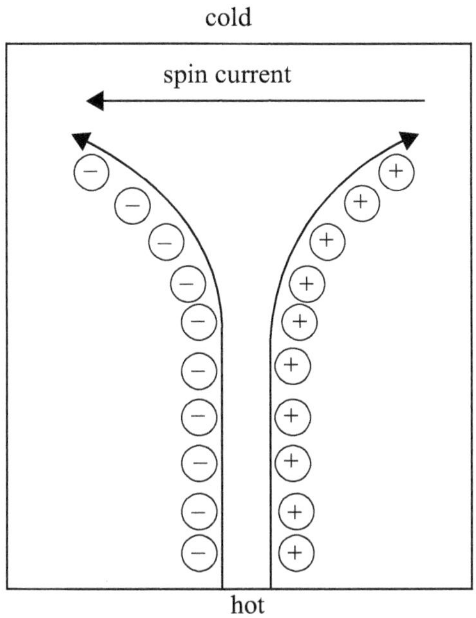

Figure 9.3. Spin Nernst effect in the antiferromagnetic honeycomb lattice.

$$Q = \frac{1}{4\pi}\int dxdy\vec{m}\cdot(\partial_x\vec{m}\times\partial_y\vec{m}). \tag{9.75}$$

The winding number measures the number of times the sphere of magnetization directions is covered upon integrating over space and must take on integer values. Here $\hat{m} = \vec{m}(x, y, z)$ is the normalized magnetization in the real space. The skyrmion number can be used to distinguish the topology of different types of spin textures. Topological charge is an invariant indicating that the trivial structure (for example, ferromagnetic state) cannot continuously deform into a topological spin texture [31]. Microscopically, the skyrmion magnetization modulation is caused by the Dzyaloshinskii–Moriya interaction. In the case of a chiral skyrmion, only one chirality is found. In the case of multiple chiralities, we have bubble skyrmions.

The skyrmion number of each configuration also characterizes the number of magnetic monopoles therein. The corresponding structure on the unit sphere is similar to a magnetic vortex, and therefore, this type of skyrmion is also referred to as a vortex-like skyrmion. Magnetic vortices have half of the topological skyrmion number $Q = \pm 1/2$. This can be determined by projecting each spin of magnetic vortices onto the unit sphere, correspondingly yielding half of the skyrmion number.

A skyrmion is described by the Hamiltonian

$$H = -J\sum_{<i,j>}\vec{S}_i\cdot\vec{S}_j + \sum_{<i,j>}\vec{D}_{ij}\cdot(\vec{S}_i\times\vec{S}_j) - g\mu_B\sum_i\vec{B}\cdot\vec{S}_i. \tag{9.76}$$

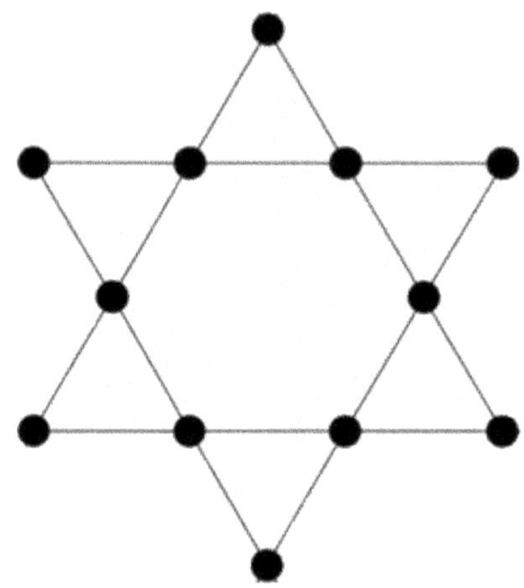

Figure 9.4. The Kagome lattice.

The competition of the terms leads to a complex phase diagram as a function of temperature and magnetic field and the regime where the skyrmion appears can be very narrow.

An interesting phenomenon is the skyrmion Hall effect. Analogous to the charge Hall effect, which describes the transverse deflection of electric charges due to the Lorentz force, magnetic skyrmions exhibit curved trajectories away from the direction of applied current because of their spin topology.

The antiferromagnetic skyrmion (AFS) can be visualized as two interpenetrating ferromagnetic skyrmions. Numerical simulations show that the AFS moves faster than ferromagnetic skyrmions and can be driven with low current densities. The Magnus force which deflects ferromagnetic skyrmions has the opposite sign for the two magnetic sublattices of an AFC.

9.9 Exercises

9.1. Find the eigenvalues given by equation (9.12).

9.2. Calculate M_k given by equation (9.9) for the ferromagnetic honeycomb lattice using the linear spin-wave formalism. Find the expression for the Berry curvature for this lattice.

9.3. Find the eigenvalues and eigenvectors of the Lieb lattice using the linear spin-wave formalism.

9.4. The Kagome lattice is shown in the figure below (figure 9.4).

Find the eigenvalues and eigenvectors for the ferromagnetic kagome lattice, using the linear spin-wave formalism (see [12]).

References

[1] Phillips P 2003 *Advanced Solid State Physics* (Boulder, CO: Westview Press)

[2] von Klitzing K, Dorda G and Pepper M 1980 New method for high-accuracy determination of the fine-structure constant based on quantized Hall resistance *Phys. Rev. Lett.* **45** 494–97

[3] Pires A S T 2019 *A Brief Introduction to Differential Geometry and Topology in Condensed Matter* (San Rafael, CA: Morgan & Claypool Publishers)

[4] Aharonov Y and Casher A 1984 Topological quantum effects for neutral particles *Phys. Rev. Lett.* **53** 319–21

[5] Meyer F and Loss D 2003 Magnetization transport and quantized spin conductance *Phys. Rev. Lett.* **90** 167204

[6] Katsura H, Nagaosa N and Balatsky A V 2005 Spin current and magnetoelectric effect in noncollinear magnets *Phys. Rev. Lett.* **95** 057205

[7] Katsura H, Nagaosa N and Lee P A 2010 Theory of thermal Hall effect in quantum magnets *Phys. Rev. Lett.* **104** 066403

[8] Nakata K, Klinovaja J and Loss D 2017 Magnonic quantum Hall effect and Wiedemann-Franz law *Phys. Rev.* B **95** 125429

[9] Nakata K, Kim S K, Klinovaja J and Loss D 2017 Magnonic topological insulators in antiferromagnets *Phys. Rev.* B **96** 224414

[10] Matsumoto R and Muramaki S 2011 Rotational motion of magnons and thermal Hall effect *Phys. Rev.* B **84** 184406

[11] Matsumoto R, Shindou R and Murakami S 2014 Thermal Hall effect of magnons in magnets with dipolar interactions *Phys. Rev.* B **89** 054420

[12] Mook A, Henk J and Mertig I 2014 Magnon Hall effect and topology in kagome lattices: a theoretical investigation *Phys. Rev.* B **89** 134409

[13] Onose Y, Ideue T, Katsura H, Shioni Y, Nagaosa N and Tokura Y 2010 Observation of the magnon Hall effect *Science* **329** 297–99

[14] Maki M and Uhrig G S 2019 Topological magnon bands for magnonics *Phys. Rev.* B **99** 174412

[15] He Y, Moore J and Varma C M 2012 Berry phase and anomalous Hall effect in a three-orbital tight-binding Hamiltonian *Phys. Rev.* B **85** 155106

[16] Shindou R, Matsumoto R, Murakami S and Ohe J 2013 Topological chiral magnonic edge mode in a magnonic crystal *Phys. Rev.* B **87** 174427

[17] Han J H and Lee H 2017 Spin Chirality and Hall-like transport phenomena of spin excitations *J. Phys. Soc. Jpn.* **86** 011007

[18] Kim S K, Ochoa H, Zarzuela R and Tserkovnyak Y 2016 A realization of the Haldane-Kane-Merle model in a system of localized spins *Phys. Rev. Lett.* **117** 227201

[19] Owerre S A 2016 A first theoretical realization of honeycomb topological magnon insulator *J. Phys.: Condens. Matter* **28** 386001

[20] Sil A and Ghosh A K 2020 First and second-order topological phases on ferromagnetic breathing Kagome lattice *J. Phys.: Condens. Matter* **32** 205601

[21] Cao X, Chen K and He D 2015 Magnon Hall effect on the Lieb lattice *J. Phys.: Condens. Matter* **27** 166003

[22] Leykam D, Andreanob A and Flach S 2018 Artificial flat band systems: from lattice models to experiments *Adv. Phys.* X **3** 1473052

[23] Konig M, Buhmann H, Molenkamp L W, Hughes T L, Liu C X, Qi X L and Zhang S C 2008 The quantum spin Hall effect: theory and experiment *J. Phys. Soc. Jpn.* **77** 03007

[24] Cheng R, Okamoto S and Xiao D 2016 Spin Nernst effect of magnons in collinear antiferromagnets *Phys. Rev. Lett.* **117** 217202

[25] Zyuzin V A and Kovalev A A 2016 Magnon spin Nernst effect in antiferromagnets *Phys. Rev. Lett.* **117** 217203

[26] Owerre S A 2017 Noncollinear antiferromagnetic Haldane magnon insulator *J. Appl. Phys.* **121** 223904
[27] Shiomi Y, Takashima R and Saitoh E 2017 Signature of magnon Nernst in an antiferromagnetic insulator *Phys. Rev.* B **96** 134425
[28] Pires A S T 2019 Topological magnons in the antiferromagnetic checkerboard lattice *Physica* E **118** 113899
[29] Sitte K E, Masell J, Reeve R M and Klaui M 2018 Perspective: magnetic skyrmions-overview of recent progress in an active research field *J. Appl. Phys.* **124** 240901
[30] Jiang W, Chen G, Liu K, Zang J, EVeltus S G and Hoffmann A 2017 Skyrmions in magnetic multilayers *Phys. Rep.* **704** 1–49
[31] Li Z X, Cao Y and Yan P 2020 Topological insulators and semimetals in classical magnetic systems arXiv:2011.09751

IOP Publishing

Theoretical Tools for Spin Models in Magnetic Systems

Antonio Sergio Teixeira Pires

Chapter 10

Topological spin liquids

As was mentioned in chapter 5, spin liquids are systems that do not break any symmetries at zero temperatures. Although they do not possess long-range order, they have exotic order, which is not described by a local order parameter. Here I will treat topological spin liquids, where, in the thermodynamic limit, the ground state is disordered and degenerate, and the degeneracy depends on the topology of the system. The simplest and best understood example of a topologically ordered system is a spin liquid with a Z_2 topological order on a two-dimensional square lattice. They are gapped, and as we shall see, with two kinds of well-defined excitations, spin ½ spinons and spin 0 excitations called visons.

Let us start with the valence bond solid (VBS) shown in figure 5.2. The bonds between four neighboring sites can fluctuate between two possible positions, as shown in figure 10.4. These resonances can cause the melting of the solid to another state with no long-range correlations of any kind: the resonant valence bond liquid mentioned in chapter 5. Note that the Heisenberg Hamiltonian does not present any implicit long-range interaction. In chapter 5, we saw that the Schwinger boson formalism in the mean-field approximation has two kinds of solutions. In the first kind, the Schwinger bosons condense in a particular mode that spontaneously breaks the SU(2) symmetry and, in the case of an antiferromagnet, leads to the Néel phase with long-range spin–spin correlations. The second kind corresponds to a magnetically disordered phase that could describe a spin liquid at zero temperature. In this phase, we have spinons' excitations that carry a spin ½. The question is whether the free spinons in the mean-field approximation are an artifact of the mean-field approximation or if they remain free in some Heisenberg model. In the first case, the inclusion of the fluctuations that were neglected could confine the spinons, would change the nature of the ground state, and the mean-field approximation would be incorrect. However, if the spinons remain deconfined, even in the presence of fluctuations, the mean-field approximation can be used, at least as a starting point. One way to include fluctuations is to use the large N

doi:10.1088/978-0-7503-3879-0ch10

formulation of the Schwinger boson theory. This formulation leads to a Z_2 gauge theory that is not directly related to the original Heisenberg model but can provide a simplified and phenomenological description of topological spin liquids. This theory will be presented in the next section at zero temperature. I will follow mainly references [1–4].

10.1 Z_2 gauge theory

I start with the Ising model in a transverse field with nearest-neighbor exchange interaction J on a square lattice described by the Hamiltonian

$$H = -J\sum_{\langle ij\rangle}\tau_i^z\tau_j^z - K\sum_i \tau_i^x, \tag{10.1}$$

where τ_i^z and τ_i^x are Pauli matrices. We have $\tau_i^x\tau_i^x = 1$, $\tau_i^z\tau_i^z = 1$, and $\tau_i^z\tau_i^x = -\tau_i^x\tau_i^z$, and

$$\tau^z = \begin{pmatrix} 1 & 0 \\ 0 & -1 \end{pmatrix}, \quad \tau^x = \begin{pmatrix} 0 & 1 \\ 1 & 0 \end{pmatrix}. \tag{10.2}$$

Let the spin-up at site i, be represented by the vector $\begin{pmatrix} 1 \\ 0 \end{pmatrix}$, and the spin-down by $\begin{pmatrix} 0 \\ 1 \end{pmatrix}$. Then τ_i^x is a spin-flip operator for site i, as we can see using (10.2)

$$\tau_i^x\begin{pmatrix} 1 \\ 0 \end{pmatrix} = \begin{pmatrix} 0 \\ 1 \end{pmatrix}, \quad \tau_i^x\begin{pmatrix} 0 \\ 1 \end{pmatrix} = \begin{pmatrix} 1 \\ 0 \end{pmatrix}. \tag{10.3}$$

When $K = 0$, the model reduces to the classical Ising model with ferromagnetic exchange interaction J, and has ferromagnetic order in the ground state with $\tau_i^z = +1$ (or -1) on every site. If $K \ll J$ we still have ferromagnetic order. On the other side, if $K \gg J$ all spins will point in the x-direction and we have $\langle\tau_i^z\rangle = 0$. A quantum-phase transition is expected to take place when $J \approx K$. We have a global Z_2 symmetry under $\tau_i^z \to -\tau_i^z$.

Now I present a duality transformation between the quantum Ising model and a dual gauge theory with a local Z_2 symmetry. Duality relations are exact relations that map a lattice system to another. A dual lattice is a lattice that lives at the center of the original lattice. I start with the one-dimensional model. Sites on the original lattice will be associated with links on the dual lattice and vice versa, as shown in figure 10.1.

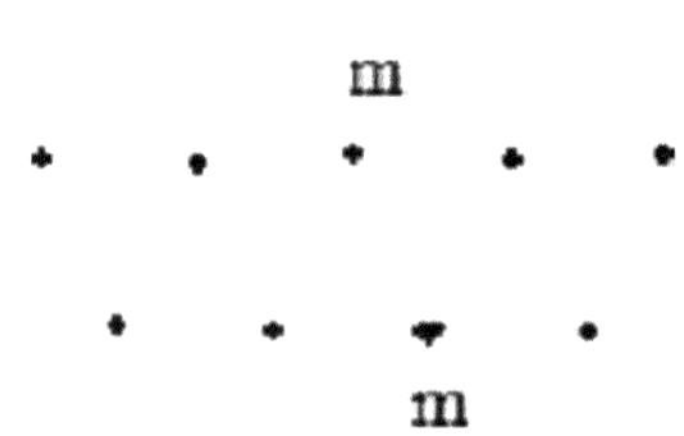

Figure 10.1. Dual transformation for a one-dimensional chain.

To construct the duality transformation, we write τ_x and τ_z in terms of the dual gauge fields σ_{ij}^{α}, which are Pauli matrices defined on the links of the dual square lattice as

$$\sigma_n^x = \tau_{n+1}^z \tau_n^z, \quad \sigma_n^z = \prod_{m<n} \tau_m^x. \tag{10.4}$$

One can check that σ_n^x and σ_m^z satisfy the same algebra as τ_n^x and τ_m^z. We have

$$\sigma_n^z \sigma_{n+1}^z = (\tau_1^x \tau_2^x \ \dots \ \tau_n^x)(\tau_1^x \tau_2^x \ \dots \ \tau_n^x \tau_{n+1}^x) = \tau_{n+1}^x, \tag{10.5}$$

since $(\tau_n^x)^2 = 1$. The Hamiltonian becomes in the dual lattice

$$H = -J \sum_n \sigma_n^x - K \sum_n \sigma_n^z \sigma_{n+1}^z. \tag{10.6}$$

Now we turn to the two-dimensional square lattice. We use the inverse relation and define

$$\tau_i^x = \prod_{pl(i)} \sigma_{jl}^z, \tag{10.7}$$

where the product is taken around an elementary plaquette with four sides on the dual square lattice, which involves the site i of the original lattice (figure 10.2) (note that σ_{ij}^{μ} are defined in the bond ij). That is, in each link of the dual lattice we associate a number 1 or −1. The product can only be 1 or −1. Next, we take

$$\tau_i^z = \prod_{jl=i}^{\infty} \sigma_{jl}^x, \tag{10.8}$$

where the product involves an infinite string that connects sites of the original direct lattice, starting from the site τ_i^z and going to infinity. For every bond of the dual lattice, which is bisected by this string, a factor σ_{ij}^x is present in the product. To be sure that the definition (10.8) is independent of the path taken by the string, we

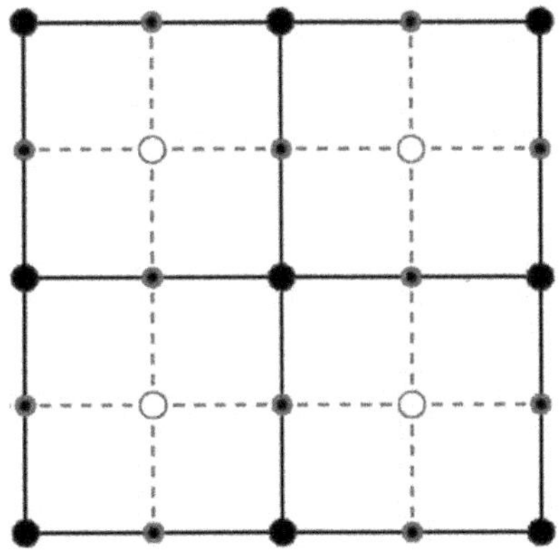

Figure 10.2. The solid lines denote the direct lattice, the dashed lines the dual lattice.

impose the constraint that the product of σ_{ij}^{x} on all bonds connected to each site on the dual square lattice is equal to one, that is, we write

$$G_i = \prod_{j\in i} \sigma_{ij}^{x} = 1, \tag{10.9}$$

where j labels the nearest neighbor of i. In terms of the dual fields, we write the Hamiltonian as

$$H = -K\sum_{\langle ij\rangle}\prod_{pl} \sigma_{ij}^{z} - J\sum_{\langle ij\rangle}\sigma_{ij}^{x}. \tag{10.10}$$

We take the product in the first term around the square plaquettes of the dual square lattice, which surrounds the sites of the original lattice. These products are a measure of the flux in the dual gauge field. If $\prod_{pl}\sigma^{z} = -1$, we say that we have a flux. If the product is 1 the flux is zero. It is easy to see that the term in J has only one σ_{ij}^{x}: we have products from i to infinite and products from j to infinite. The non-common factor is the term σ_{ij}^{x} connecting i to j.

As was said before, for $J \gg K$, the original Ising model has ferromagnetic order along the positive z-direction. In this case, the dual gauge theory has a ground state with $\sigma_{ij}^{x} = 1$ for all links. It is known that the low energy excitations of the original Ising model are droplets of spins $\tau^{z} = -1$ in the background of spins $\tau^{z} = 1$, as shown in figure 10.3 [5]. Such droplets are created from the ferromagnetic ordered state by flipping all the spins inside them. We can therefore write

$$|\text{drop}\rangle = \prod_{i\in\text{drop}} \tau_i^{x} \,|\, 0\rangle, \tag{10.11}$$

where $|0\rangle$ is the ground state (ferromagnetically ordered). We can also write

$$|\text{drop}\rangle = \prod_{\langle ij\rangle\in c} \sigma_{ij}^{z}, \tag{10.12}$$

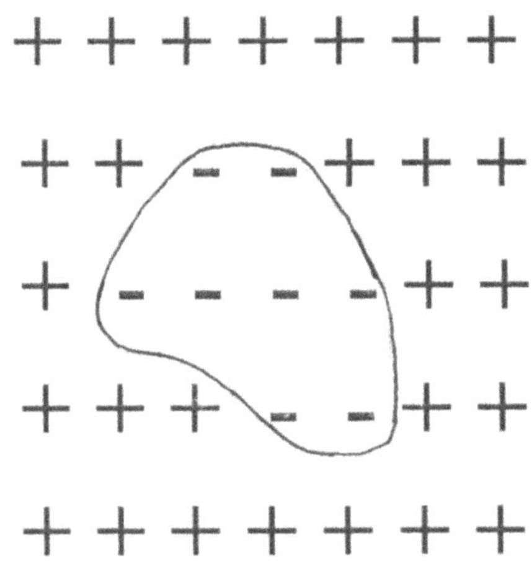

Figure 10.3. A block of flipped spins. A domain wall is a closed path that surrounds the droplet. The pluses (minuses) denote spins pointing up (down).

where c is the closed path surrounding the droplet. This phase is called the *confining* phase of the gauge theory. Let us consider two excitations in the dual lattice with value -1:

$$\prod_{l\in i} \sigma_{il}^{x}, \prod_{l\in j} \sigma_{jl}^{x}. \tag{10.13}$$

We call these excitations the Z_2 charges. Charges on the sites i and j of the dual lattice demand energy linearly depending on their separation.

We turn now to the paramagnetic phase of the original Ising model in the limit $K \gg J$. The ground state of the Ising model corresponds to all spins with $\tau_i^x = 1$. The ground state of the dual theory is given by $\prod_{pl} \sigma_{ij}^z = 1$ for all plaquettes. Excited states are given by simple plaquettes with $\prod_{pl} \sigma_{ij}^z = -1$. This excitation is called ***vison*** (or a Z_2 vortex). A vison can be created by acting on the ground state with τ_i^z, where i is the site of the original lattice, which is in the center of the corresponding dual plaquette. Hence, a vison consists of a site with $\tau_i^x = -1$ in the original lattice. A vison is a nonmagnetic excitation with spin $S = 0$. Two visons on the same plaquette annihilate, and they can be created only in pairs.

10.2 Dimers

Now we consider a lattice with dimers. We take $\sigma_{ij}^x = 1$ in a link if there is no dimer in the bond, and $\sigma_{ij}^x = -1$ if there is a dimer. This model is not the Z_2 model discussed in section 10.1, where we could have, for instance, all sites with $\sigma_{ij}^x = 1$. Note that here we have changed the roles of σ^x with σ^z. The interactions on the lattice must be invariant under the Z_2 gauge transformation $\sigma_{ij}^z \to \eta_i \sigma_{ij}^z \eta_j$, where $\eta_i = \pm 1$ may take an arbitrary value at each lattice site. We define an operator (called the flux operator) on the four sites around a plaquette as

$$F_p = \sigma_{ij}^z \sigma_{jk}^z \sigma_{kl}^z \sigma_{li}^z, \tag{10.14}$$

where σ_{ij}^z change the sign of σ_{ij}^x. This product is invariant under the gauge transformation mentioned above. The action of F_p in a pair of dimers is shown in figure 10.4.

The Hamiltonian, which describes the resonance valence bonds, is thus given by

$$H = -K \sum_{p(ijkl)} F_p, \tag{10.15}$$

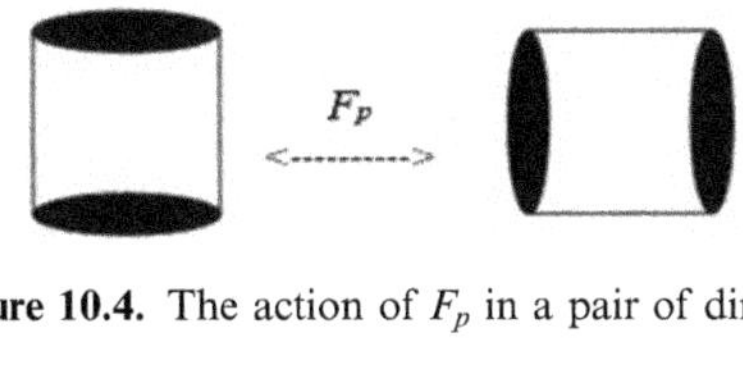

Figure 10.4. The action of F_p in a pair of dimers.

where $F_p = \pm 1$. Each spin can form only a singlet, so that only one valence bond is present at each site. We define a gauge charge Q_i for each site as

$$Q_i = \prod_{j=1}^{4} \sigma_{ij}^x, \tag{10.16}$$

and we impose the condition that $Q_i = -1$. The Hilbert space is obtained using the projection operator

$$P = \prod_i \frac{1}{2}(1 - Q_i). \tag{10.17}$$

But there is one detail. The number of dimers at each site can be one or three. To avoid the possibility of three bonds out of each site we add a term $-J\sigma_{ij}^x$ (where J is the gain in energy of having no dimer on a given link) to the Hamiltonian, which minimizes the number of dimers at each site, and energetically selects one valence bond only. The Hamiltonian is then written as

$$H = -K \sum_{p(ijkl)} \sigma_{ij}^z \sigma_{jk}^z \sigma_{kl}^z \sigma_{li}^z - J \sum_{ij} \sigma_{ij}^x, \tag{10.18}$$

with the constraint $Q_i = 1$.

In the ground state, for $J = 0$, we have $F_p = 1$ in all plaquettes (since this value minimizes the first term in 10.18), but σ_{ij}^x can fluctuate since

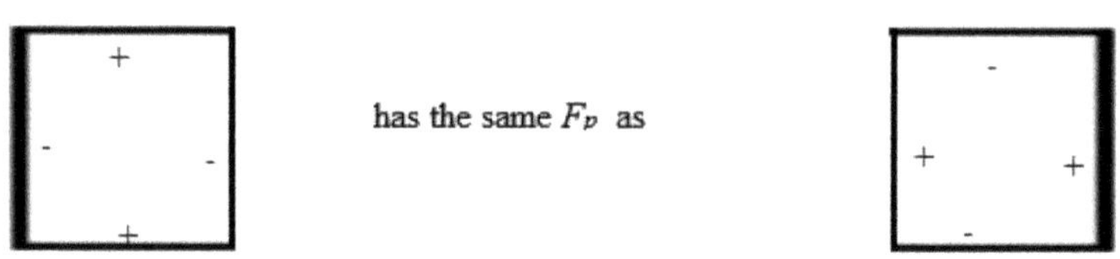

The dimers are disordered, and as we saw before, the excitations are visons, which have an energy cost of the order of K. We have seen that this state is deconfined. It corresponds to the RVB (resonance valence bond) where the valence bonds fluctuate.

Let us consider the case $K \ll J$ (the VBS (valence bond solid) phase). The energy would be a minimum if we had +1 in all links (no dimer). But we have N_d dimers on the lattice. We must minimize the energy with this constraint. The dimers are ordered. The bonds are kept fixed, and the plaquette flux fluctuates (the visons are condensed), and the spinons are confined. A lattice, as shown in figure 5.2, tends to remain as such. At $J = 0$, the energy of two visons is independent of their separation. Away from $J = 0$ visons acquire a finite bandwidth and non-trivial short-range interactions.

Up to now, I have talked only about the gauge field. Let us consider the interaction between spins in a singlet. The break of a valence bond results in two spinon excitations, which corresponds to a spin-flip excitation with spin-one (breaking a valence bond cost energy of order J). A spinon arises when one valence bond is broken, thereby creating two dangling bonds with $\pm 1/2$ spins. They do not carry an electric charge, since breaking a valence bond does not add any charge to the system. As they happen to be topological, it is convenient to introduce a topological charge to them. In the VBS case, when we separate the spinons, we create a mismatch in the array of dimers, and the cost in energy increases linearly with the distance between the spinons (each link with $\sigma_{ij}^{x} = -1$ introduces a cost in energy). Therefore, they are confined in a VBS (figure 10.5).

In the RVB case, as shown in figure 10.6, we have a valence bond disordered state. Energy cost stays finite when spinons are separated, and the spinons are deconfined. In the deconfined phase, as we saw before, we have $F_p = 1$, and the bond fields fluctuate (the dimers are disordered).

In the VBS phase, the bond field is kept fixed. The dimers are ordered, and the visons are condensed. There is no connecting valence in a site with a spinon. If we destroy a dimer, the flux in the plaquette changes, i. e., we create a vison. As was said before, we have a gauge charge of -1 at the site, and we attribute a charge $Q_i = -1$ to the spinon. When the spin is displaced in one VBS the flux changes its value along the path. In other words, it fluctuates.

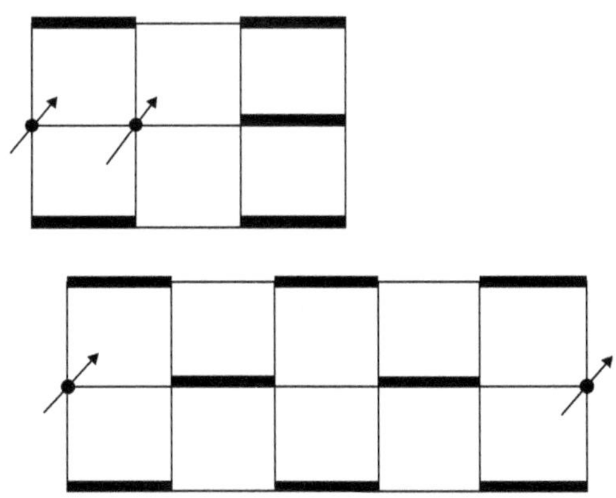

Figure 10.5. Two spinons are excited by breaking a valence bond.

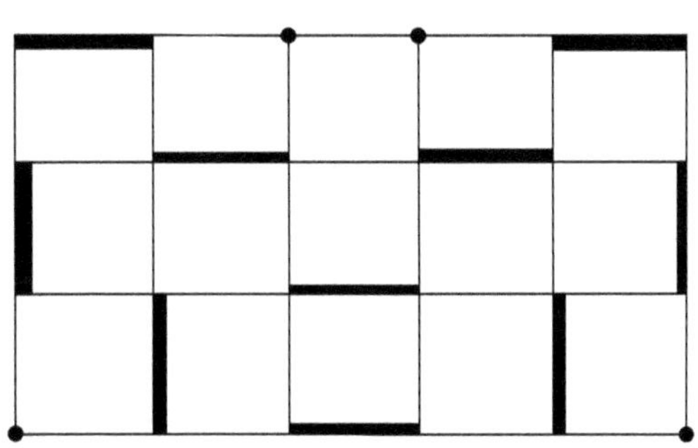

Figure 10.6. Resonance valence bond configuration.

We can introduce a spinon creation operator at each site with spin up/down: $b_{i\alpha}^{\dagger}$. The full Hamiltonian is then written as

$$H = -K\sum_{p(ijkl)}\sigma_{ij}^{z}\sigma_{jk}^{z}\sigma_{kl}^{z}\sigma_{li}^{z} - J\sum_{ij}\sigma_{ij}^{x} - t_s\sum_{\langle ij\rangle}\sum_{\alpha}\sigma_{ij}^{z}b_{i\alpha}^{\dagger}b_{j\alpha} + \Delta\sum_{i\alpha}b_{i\alpha}^{\dagger}b_{i\alpha} + V\sum_{i}\left[\left(\sum_{\alpha}b_{i\alpha}^{\dagger}b_{i\alpha} - \frac{1}{2}\right)^{2} - \frac{1}{4}\right]. \tag{10.19}$$

The third term represents spinon ‘hopping’ and their interaction with the Z_2 gauge field, the fourth term represents the energy cost to create a spinon and the last term include a substantial penalty V when more than one spinon is present on the same site.

It is common in the literature to make an analogy with the electromagnetic theory, where a vison is thought of as a ‘magnetic’ flux and σ_{ij}^{x} as an electric field.

The other class of spin liquid, named *critical spin liquids*, is more complicated than topological spin liquids. They can be understood in terms of valence bonds between spins separated by arbitrary distances. The spinon excitations are gapless, strongly interacting with the valence bond background, and cannot be considered free particles (for more details, see [1]).

References

[1] Alet F, Walczak A M and Fisher M P A 2006 Exotic quantum phases and transitions in correlated matter *Physica* A **369** 122–42

[2] Fisher M P A 2004 Duality in low dimensional quantum field theories *Strong Interactions in Low Dimensions* ed D Baeriswyl and L Degiorgio (Dordrecht: Springer) pp 419–38

[3] Misguich G 2008 Quantum Spin Liquids arXiv:0809.2257

[4] Misguich G and Lhuillier C 2005 Two dimensional quantum antiferromagnets *Frustrated Spin Systems* ed H T Diep (Singapore: World Scientific)

[5] Kogut J B 1979 An introduction to lattice gauge theory and spin systems *Rev. Mod. Phys.* **51** 659–713

IOP Publishing

Theoretical Tools for Spin Models in Magnetic Systems

Antonio Sergio Teixeira Pires

Chapter 11

Numerical methods for spin models

Computational approaches have become one of the central cornerstones in condensed matter alongside theoretical and experimental approaches. In this chapter, I will provide an overview of numerical methods. Since it is not possible to go into full details, I will present only the general ideas. In the other chapters, I have tried to explain the material in a self-consistent way. That is not the case with this chapter, where the idea is only to familiarize the reader with the methods. For more details, I refer the reader to the references cited.

11.1 Monte Carlo

Monte Carlo is not a specific technique to solve problems. Indeed, it is a common name for several methods based on sampling and statistical physics, which is used to estimate an event's possible outcomes. The basis for Monte Carlo simulations, which allows us to build a stochastic model by the method of statistical trials, is the 'Law of Large Numbers':

Let $\{x_1, x_2 \ldots x_N\}$ be a sequence of independent and identically distributed random variables with mathematical expectation $E(x)$. Then, the average $\frac{1}{N}\sum_{n=1}^{N} x_n$ converges to $E(x)$ as $N \to \infty$.

Because of its simplicity, the development of fast computers, and new algorithms, Monte Carlo methods have become more powerful and popular to investigate problems in a wide range of areas when the number of degrees of freedom involved makes the problem unamenable using any other approach. In general, a Monte Carlo experiment follows the basic steps:

1. Define the model.
2. Identify the probability distribution of the income variable.
3. Run independent simulation experiments several times to cover a representative number of outcomes.

In equilibrium statistical mechanics, one wants to determine the behavior of a large system in thermodynamic equilibrium. Averages of an operator A is obtained by

$$\langle A \rangle = \frac{1}{Z} tr(Ae^{-\beta H}) = \frac{1}{Z} \sum_{n} \langle n | Ae^{-\beta H} | n \rangle, \tag{11.1}$$

where H is the Hamiltonian of the system and $Z = tr(e^{-\beta H})$ is the partition function. In the case of a classical system, equation (11.1) simplifies since all operators commute. In such a case, the averages are the sum over all possible configurations $\{x\}$ of the system.

$$\langle A \rangle = \frac{1}{Z} \sum_{\{x\}} A(\{x\}) e^{-\beta E\{x\}}. \tag{11.2}$$

It is essential to make the difference between 'Classical Monte Carlo' (CMC) and 'Quantum Monte Carlo' (QMC) since QMC approaches differ significantly from Monte Carlo methods for classical systems. The QMC covers many stochastic methodologies developed to determine ground state, excited states, or finite-temperature equilibrium properties of various quantum systems. Although many of the ideas applied to CMC can be adapted to QMC, it is vital to treat them separately.

11.2 Classical Monte Carlo

The Monte Carlo technique is a probabilistic approach to simulate a thermal ensemble applicable to a large class of statistical physics problems. An ideal statistical ensemble consists of an infinite number of copies of a system intending to take thermal averages. We cannot do that on a computer with finite memory. Therefore, we generate the ensemble members one at a time and replace the ensemble average with a time average. The accuracy of a computation is so limited by computing time instead of computer memory [1].

The equilibrium probability distribution of the variable x is $p_\beta(x) = e^{-\beta E(x)}/Z$. Now, our task is to obtain $A(\{x\})$ that follows the distribution $p_\beta(x)$. Fortunately, it is possible to build a Markov chain to get an equilibrium configuration starting from an arbitrary one by successively applying a transition matrix $p(\{x\} \to \{x'\})$. Once a chain has reached an equilibrium distribution, it remains there. In short, the process is as follows:

1. Choose an arbitrary initial configuration $\{x\}$.
2. Make a trial change $\{x\} \to \{x'\}$.
3. Accept the change with probability $p = \min\begin{cases} 1 \\ e^{-\beta[E(\{x\}) - E(\{x'\})]} \end{cases}$.
4. Go back to 1 until the system is in equilibrium.

This simple process is known as the Metropolis algorithm. It is a local algorithm because only local variables are updated in each Monte Carlo step. That imposes a severe limitation when dealing with systems close to a phase transition or at $T \approx 0$.

The equilibration time becomes prohibitive due to the large correlations developed in the system. Many non-local algorithms were developed to overcome this problem. See, for instance, Wolff [2] and Evertz [3]. In 2001, a powerful technique developed by Wang and Landau [4] was proposed. It starts by rewriting the partition function as

$$Z = \sum_E g(E)e^{-\beta E}, \tag{11.3}$$

where $g(E)$ stands for the density of states. Instead of sampling the probability distribution at a fixed temperature, a random walk is performed in energy space to estimate $g(E)$ directly. It is particularly efficient when the free energy of the system has a rough landscape with multiple competing local minima.

The Wang–Landau sampling performs a random walk-in energy space by randomly making trial changes in the system. The probability of visiting a given energy level, E, is proportional to the reciprocal density of states. The number of visits at each energy level is controlled by an auxiliary histogram $H(E)$. The estimated density of states is modified in a systematic way to produce a 'flat' histogram over the allowed range of energy. At the beginning of the process $g(E)$ is not known; all entries are set to $g(E) = 1$ for all possible energy levels. Then, a random walk-in energy space is done, moving from one energy level to another with probability proportional to $1/g(E)$. The new configuration is accepted with probability

$$p(E_1 \to E_2) = \min\begin{cases} 1 \\ g(E_1)/g(E_2)' \end{cases} \tag{11.4}$$

where E_1 and E_2 are the energies before and after the move, respectively. Each time an energy level E is visited, the current value is multiplied by a modification factor $f > 1$, and the correspondent entry of the auxiliary histogram is updated $H(E) \to H(E) + 1$ [4]. The process continues until a 'flat' histogram $H(E)$ is obtained. Each time a flat histogram is obtained, $H(E)$ is set to zero, the modification factor is reduced as $f \to \sqrt{f}$ and the process starts over until f is reduced to a previous set value.

11.3 Quantum Monte Carlo

Because the eigenstates of the quantum Hamiltonian H are generally not known, it is not obvious how to calculate the partition function

$$Z = tr(e^{-\beta H}), \tag{11.5}$$

at finite temperature using QMC methods. We have to resort to map the quantum model to a representation with a probabilistic interpretation. I will show below the stochastic series expansion (SSE) approach, a highly efficient and easily implementable method [5].

One convenient way to do that is to use the path-integral method to transform the trace of the partition function into a form that can be sampled using Monte Carlo methods. We start by writing the term $e^{-\beta H}$ as a product of L operators with $\Delta_\tau = \beta/L$ (called a 'time slice') in the exponent:

$$Z = tr\left\{\prod_{l=1}^{L} e^{-\Delta_\tau H}\right\}. \tag{11.6}$$

We insert a complete set of states, in a suitably chosen basis, between each of the exponentials and write the partition function as

$$Z = \sum_{\alpha_0}\sum_{\alpha_1}\dots\sum_{L-1}\langle\alpha_0|e^{-\Delta_\tau H}|\alpha_{L-1}\rangle\dots\langle\alpha_2|e^{-\Delta_\tau H}|\alpha_1\rangle\langle\alpha_1|e^{-\Delta_\tau H}|\alpha_0\rangle. \tag{11.7}$$

As we saw in chapter 7, the exponential operator is equivalent to the time evolution operator with imaginary time $t = -i\Delta_\tau$. The index l corresponds to a discrete set of imaginary time points $\tau_l = l\Delta_\tau$, with $0 \leqslant \tau_l \leqslant \beta$ and periodic time-boundary conditions $\alpha_L = \alpha_0$ (to keep the trace non-zero). Because the time step Δ_τ is small, the matrix elements can be computed to some approximation with a total error of order Δ_τ^p, where p depends on the approximation used.

The powers of the Hamiltonian can be decomposed into strings of two-body operators, $H_{b_n}\dots H_{b_2}H_{b_1}$, with all combinations of the operators

$$H_b = H_{i(b),j(b)}, \quad b \in \{1, \dots, N_b\}, \tag{11.8}$$

on a finite lattice in any number of space dimensions D (where $N_b = DL$ for nearest-neighbor interactions on a D-dimensional simple cubic lattice). Denoting a string of n operator indices $b_1, b_2, \dots, b_n$ by S_n, the partition function becomes

$$Z = \sum_{n=0}^{\infty}\frac{(-\beta)^n}{n!}\sum_{S_n} tr\{H_{b_n}\dots H_{b_2}H_{b_1}\}, \tag{11.9}$$

where the traces of the strings of permutation operators are positive definite. With a simple scheme to evaluate the traces in terms of permutation cycles, the operator strings can be sampled using a Metropolis Monte Carlo method, and various thermodynamic expectation values can be evaluated.

Note that it is not necessary to compute the traces analytically. They also can be sampled on a chosen basis along with operator strings. The sampling weight in the extended space comprising states and operator strings has matrix elements

$$\langle\alpha|H_{b_n}\dots H_{b_2}H_{b_1}|\alpha\rangle,$$

in place of the full traces in equation (11.9). We have then

$$Z = \sum_{n=0}^{\infty}\sum_{S_n}\sum_{\alpha}\frac{(-\beta)^n}{n!}\langle\alpha|H_{b_n}\dots H_{b_2}H_{b_1}|\alpha\rangle. \tag{11.10}$$

If the weight $(\beta^n/n!)\langle\alpha|H_{b_n}\dots H_{b_2}H_{b_1}|\alpha\rangle$ is non-negative for all α, n and S_n, one has obtained a configuration space that can be interpreted probabilistically, allowing Monte Carlo simulations.

The 'diagonal' terms either insert or remove diagonal bonds in the operator string S_n, thereby changing the current expansion order n. The 'off-diagonal' terms can modify the state $|\alpha\rangle$ and exchange diagonal into off-diagonal bond terms while conserving n. Since at order n, one has a factor $(-\beta)^n$, the matrix elements must be negative if the total weight is to be positive for all n, so that each matrix element can be interpreted as a probability for use in a Metropolis scheme. The positive diagonal matrix elements of H can, in principle, always be made negative by subtracting a suitable constant. The situation is more complicated if there are positive off-diagonal matrix elements. This is a manifestation of the so-called 'sign problem'. It can essentially be avoided only when all terms in the Hamiltonian have a negative sign in front of them. That is the case for the ferromagnet. In the case of an antiferromagnet on a bipartite lattice, we can make all the terms have a negative sign through a basis rotation. However, the QMC method cannot be applied to all quantum mechanical problems on frustrated systems.

One way to circumvent the sign problem is by formally assigning the sign of a configuration on the observables calculated, and the true expectation value of the observable recovered by dividing by the average sign. However, the procedure leads to large errors for low-temperatures or large system sizes [6].

11.4 High-temperature expansions

The formalism is, in principle, straightforward. The idea is to expand a thermodynamic function Q in the inverse temperature β, around the infinite temperature limit $\beta = 0$ [7, 8]. The high-temperature series is obtained by merely expanding the exponential function. For instance

$$Z \approx \sum_{n=0}^{\infty} \frac{(-\beta)^n}{n!} trH^n, \tag{11.11}$$

where the trace covers the full Hilbert space of the quantum system. For a variable Q, we have

$$\langle Q \rangle = \frac{trQe^{-\beta H}}{tre^{-\beta H}} = \frac{trQ\sum_{n=0}^{\infty}(-\beta H)^n/n!}{tr\sum_{n=0}^{\infty}(-\beta H)^n/n!}. \tag{11.12}$$

For a system with a finite number of discrete states, such as a spin system, we can divide the numerator and denominator by the partition function Z_0 at $T = \infty$. Z_0 gives the number of states. We define

$$\bar{Q} = \frac{trQ}{tr1}. \tag{11.13}$$

$\bar{Q}$ is hence the ensemble average of Q at infinite temperature. At finite temperatures, we have

$$
\begin{aligned}
\langle Q \rangle &= \frac{\bar{Q} - \beta\overline{QH} + \beta^2\overline{QH^2}/2! + \dots}{1 - \beta\bar{H} + \beta^2\bar{H}^2/2! + \dots} \\
&= \bar{Q} - \beta(\overline{QH} - \bar{Q}\bar{H}) + \frac{\beta^2}{2!}(\overline{QH}^2 - 2\overline{QH}\bar{H} + 2\bar{Q}\bar{H}^2 - \bar{Q}\bar{H}^2) + \dots
\end{aligned} \tag{11.14}
$$

Hence, if

$$
\langle Q \rangle = \sum_n a^n \beta^n, \tag{11.15}
$$

we find

$$
\begin{aligned}
a_0 &= \bar{Q}, \quad a_1 = -(\overline{QH} - \bar{Q}\bar{H}) \\
a_2 &= \frac{1}{2!}(\overline{QH}^2 - 2\overline{QH}\bar{H} + 2\bar{Q}\bar{H}^2 - \bar{Q}\bar{H}^2).
\end{aligned} \tag{11.16}
$$

Such series expansions can be constructed using graphical methods, and several graphical methods have been developed. The Hilbert space has infinite dimension, so we have to make finite truncation of order N and use (assuming H is composed of local interactions only) a limited interaction range of H^n-terms combined with the translational invariance of the infinite lattice.

In frustrated systems, the high-temperature series often do not extend to temperatures much lower than the exchange constant J. Sometimes it is possible to go to lower temperatures by applying series-extrapolation techniques such as Padé approximants.

11.5 The density matrix renormalization group

The density matrix renormalization group (DMRG) is one of the most potent numerical tools in condensed matter physics. It was formulated in 1992 by S.R. White to study the ground state of strongly interacting electronic systems in one dimension [9, 10], such as the Heisenberg and Bose–Hubbard models. It has also been adapted to simulate two-dimensional models and the temporal evolution of one-dimensional systems [11]. For one-dimensional systems, the DMRG gives nearly exact results. However, in higher dimensions, the best DMRG calculations do not match the results obtained from other methods.

The DMRG is a numerical iterative variational technique devised to obtain a quantum many-body system's low-energy physics. Let us consider a system consisting of sites aligned in a one-dimensional open-boundary chain. The Hilbert space grows exponentially with size. For example, a spin ½ chain of length L has 2^L degrees of freedom. In the DMRG, we systematically transform the Hilbert space of the system to one with drastically reduced effective degrees of freedom (those most important for a target space, where the target space is often the ground state) while keeping the low-energy physics as unchanged as possible.

The DMRG was inspired by the renormalization group technique introduced by Kadanoff [12] and Wilson [13]. The main ideas of the method can be understood in the following way. Suppose we want to obtain the ground state of a one-dimensional system of size L with Hamiltonian H. If d is the spin dimension, as was said before, the corresponding Hilbert space grows exponentially as d^L. Diagonalizing the complete Hamiltonian is impossible using numerical techniques. Let us introduce initially two-block spins, left (L) and right (R) blocks (figure 11.1).

The Hamiltonian for blocks L and R, (H_L and H_R) can be exactly diagonalized. The system size is augmented by introducing two spins, one in block L and the other in block R, such that the new Hamiltonian is

$$H_L' = H_L + H_S + H_{LS}. \tag{11.17}$$

And similarly, for H_R. A 'superblock' describing the total system is constructed adding H_L' and H_R'

$$H_{\text{super}}' = H_L' + H_R'. \tag{11.18}$$

The superblock is now taken as the initial configuration, one new spin is added, the Hamiltonian is diagonalized, and the process goes on until reaching the desired precision. However, as the system size grows, the Hilbert space increases exponentially. It must be truncated to a size M. The elements of this effective M-dimensional

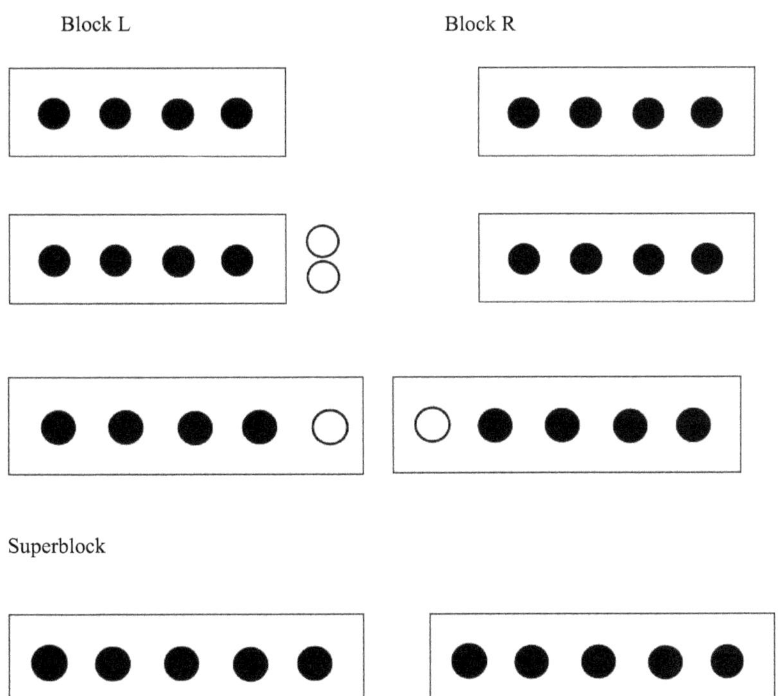

Figure 11.1. Pictorial construction of the DMRG process. The superblock is built from two blocks and two single sites. The base states of the left and right blocks are $|\alpha\rangle_L$ and $|\alpha\rangle_R$, respectively. They are made as an M-dimensional reduced Hilbert space.

Hilbert space are chosen as the most relevant states of block L. The state of the combined system can be written as

$$|\psi\rangle = \sum_{\alpha_L,\alpha_R,\lambda_L,\lambda_R} C_{\alpha_L,\alpha_R,\lambda_L,\lambda_R}|\alpha_L\rangle|\alpha_R\rangle|\lambda_L\rangle|\lambda_R\rangle. \tag{11.19}$$

Here $|\lambda_L\rangle$, $|\lambda_R\rangle$ are the basis for the spins introduced at the left and right sides, respectively. The system's ground state $|\psi\rangle_0$ and energy E_0 can be found by minimizing $E_0 = \min(\langle\psi|H_{\text{super}}|\psi\rangle)$.

Implementing the DMRG algorithm is a lengthy work. Some computational tricks, such as using the Lanczos algorithm of matrix diagonalization to obtain the ground state for the superblocks, should be used. For details of the method, see Schollwock [11], Tezuka [14] and Lauchli [6].

11.6 Exact diagonalization

The exact diagonalization (ED) technique is a powerful method that can be used to treat almost any quantum lattice model. In this method, one solves the time-independent Schrödinger equation

$$H\,|\psi\rangle = E\,|\psi\rangle, \tag{11.20}$$

of a finite quantum lattice by diagonalizing the Hamiltonian numerically without any approximation. That is feasible only for small systems since the Hilbert space grows exponentially with the system size. Generally, the calculation is performed at zero temperature, and so we obtain information about the system's low-energy physics.

The implementation of an exact diagonalization program consists of four parts:

1. Representation of the computational basis.
2. A numerical representation of the Hamiltonian matrix.
3. An algorithm for solving the eigenvalue problem.
4. A numerical procedure that calculates observables from the obtained eigenstates.

In the following, I will discuss the first three items. For a spin S, the Hilbert space has dimension $N_h = (2S + 1)^N$, where N is the number of spins, thus becoming quickly intractable, for N as small as a few tens. For instance, for $N = 30$ and $S = ½$, we have $N_h \sim 1$ billion. However, it is possible to reduce the Hilbert space's dimensions by exploiting some system symmetries or restricting them to states of a given total magnetization. For instance, if we consider the case with n spins up (and therefore N-n spins down), we have $N_h = N!/[n!(N - n)!]$, and for $N = 30$, N_h is reduced to 155 million at most. Sometimes the number of non-zero matrix elements is much smaller than the total number of matrix elements (a matrix in which most of its components are zero is called a *sparse* matrix). In conjunction with symmetry, sparseness can be used to reduce the Hamiltonian matrix into blocks of much smaller sizes.

We have now to diagonalize the $N_h \times N_h$ Hamiltonian matrix

$$H_{\mu,\nu} = \langle \mu | H | \nu \rangle, \tag{11.21}$$

where, for spin ½

$$|\mu\rangle = |\sigma_\mu(1)\rangle \otimes |\sigma_\mu(2)\rangle \otimes \ldots \otimes |\sigma_\mu(N)\rangle, \tag{11.22}$$

where $\sigma_\mu(i) = \pm 1/2$ for the ith spin and μ and ν run on all possible N_h states.

As an example, we use the Heisenberg model

$$H = \sum_{\langle ij \rangle} J\left[\frac{1}{2}(S_i^+ S_j^- + S_i^- S_j^+) + S_i^z S_j^z\right]. \tag{11.23}$$

Suppose we have only two spins ½. Using the basis $\{|\uparrow\uparrow\rangle, |\uparrow\downarrow\rangle, |\downarrow\uparrow\rangle, |\downarrow\downarrow\rangle\}$ the matrix representation is

$$\begin{pmatrix} J/4 & 0 & 0 & 0 \\ 0 & -J/4 & J/2 & 0 \\ 0 & J/2 & -J/4 & 0 \\ 0 & 0 & 0 & J/4 \end{pmatrix}. \tag{11.24}$$

The block diagonal reflects the conservation of the total magnetization. To represent states of the Hilbert space in the computer, we must choose a representation that makes it simple to act with the Hamiltonian (or operators) on the states, and to localize a given state in the basis. For example, a chain of spins ½ can be represented in binary coding as

$$\begin{aligned} |0\rangle &= |\downarrow, \downarrow, \downarrow, \ldots, \downarrow\rangle = [0\ldots000], \\ |1\rangle &= |\uparrow, \downarrow, \downarrow, \ldots, \downarrow\rangle = [0\ldots001], \\ |2\rangle &= |\downarrow, \uparrow, \downarrow, \ldots, \downarrow\rangle = [0\ldots010], \\ |3\rangle &= |\uparrow, \uparrow, \downarrow, \ldots, \downarrow\rangle = [0\ldots011], \end{aligned} \tag{11.25}$$

with $H_{ij} = \langle i | H | j \rangle$.

Diagonalization can be exact, where the answer is found exactly, or it can be iterative, where the solution is approached as the number of iterations goes to infinity. Iterative diagonalization algorithms are based on successive steps of a trial function until the required accuracy is reached. A very convenient algorithm is the *Lanczos* algorithm. We start with some initial normalized guess vector $|\phi_1\rangle$. Then we construct

$$\beta_1 |\phi_2\rangle = H|\phi_1\rangle - \alpha_1 |\phi_1\rangle, \tag{11.26}$$

$$\beta_n |\phi_{n+1}\rangle = H|\phi_n\rangle - \alpha_n |\phi_n\rangle - \beta_n |\phi_{n-1}\rangle, \tag{11.27}$$

where

$$\alpha_n = \langle \phi_n | H | \phi_n \rangle, \;\; \beta_n = \left|\left| \langle \phi_{n+1} | H | \phi_n \rangle \right|\right|. \tag{11.28}$$

The Hamiltonian has a tridiagonal form. After n steps, we have:

$$H_t = \begin{pmatrix} \alpha_1 & \beta_1 & & & & \\ \beta_1 & \alpha_2 & \beta_2 & & & \\ & \ddots & \ddots & \ddots & & \\ & & \ddots & \ddots & & \beta_{n-1} \\ & & & & \beta_{n-1} & \alpha_n \end{pmatrix}. \tag{11.29}$$

If $n = N_h$, this transformation becomes exact: $H_t = H$. Since the diagonalization of a tridiagonal matrix is a quick and easy operation, this procedure gives us a convenient numerical technique to obtain a few low-lying states of large matrices. In practice $n \sim 100$ is usually enough to reach a tolerance of 10^{-14} for the ground state energy. Note that to implement the Lanczos algorithm, one needs only $|v\rangle = H|u\rangle$ operations. The tridiagonal matrix can be diagonalized using the LAPACK routine dsterf. LAPAC is a free software package that provides routines for solving simultaneous linear equations, eigenvalue problems, etc. For a presentation of the method, see Troyer [15], Lauchli [6], Giammazzi [16], and Weisse and Fehske [17].

11.7 Coupled-cluster method

The coupled-cluster method (CCM) is a powerful and versatile numerical technique in many-body physics. The method aims to solve the ground state Schrödinger equation

$$H\,|\psi\rangle = E\,|\psi\rangle \text{ and } \langle\tilde{\psi}\,|{=}E\langle\tilde{\psi}\,|. \tag{11.30}$$

We start by choosing a normalized starting state (called the model state) $|\phi\rangle$ on top of which we incorporate later in a systematic way, the multi-spin correlation in the exact ground state $|\psi\rangle$ and $|\tilde{\psi}\rangle$. The CCM employs the exponential parametrization $|\psi\rangle = e^S|\phi\rangle$ and $\langle\tilde{\psi}| = \langle\phi|\tilde{S}e^{-S}$. We choose these states such that $\langle\tilde{\psi}|\psi\rangle = 1$. The correlation operator S is written as $S = \sum_{I\neq 0} S_I C_I^+$ and its counterpart is $\tilde{S} = 1 + \sum_{I\neq 0}\tilde{S}_I C_I^-$. The index I is a set index that stands for the set of lattices sites whose spin projections on the quantization axis are changed with respect to their values in the state $|\phi\rangle$. The correlations coefficients $(S_I, \tilde{S}_I)$ are calculated by requiring the ground state energy expectation value

$$\bar{H} = \langle\tilde{\psi}\,|H|\psi\rangle, \tag{11.31}$$

to be a minimum with respect to each of them. That yields the coupled set of equations

$$\langle\phi|C_I^- e^{-S}He^S|\phi\rangle = 0 \text{ and } \langle\tilde{\phi}|\tilde{S}(e^{-S}He^{-S} - E)C_I^+)C_I^+|\phi\rangle = 0,$$

for all $I \neq 0$, which we solve for the correlation coefficients $(S_I, \tilde{S}_I)$ within specific truncation schemes described below.

To treat each lattice site on an equal footing, we perform a mathematical rotation of the local spin axes on each lattice site, such that every spin of the model state aligns along the negative z-axis. From now on, our description of the spins is given wholly in terms of these locally-defined spin coordinate frames. The multi-spin creation operator may be written as

$$C_I^+ \equiv s_{i_1}^+ s_{i_2}^+ \dots s_{i_n}^+, \tag{11.32}$$

where $s_i^+ \equiv s_i^x + is_i^y$. Having solved the multi-spin cluster correlation coefficients $(S_I, \tilde{S}_I)$, we can then calculate the ground state energy E from the relation

$$E = \langle \phi | e^{-S} H e^S | \phi \rangle, \tag{11.33}$$

and the average ground state on-site magnetization M from the relation

$$M = -\frac{1}{N} \langle \tilde{\psi} | \sum_{i=1}^{N} s_i^z \, | \psi \rangle, \tag{11.34}$$

which holds in the rotated local spin coordinates.

Although the CCM technique is exact if a complete set of multi-spin configurations $\{I\}$ with respect to the model state $|\phi\rangle$ is included in calculating the correlation operators S and $\tilde{S}$, in practice, it is necessary to use systematic approximation schemes to truncate them to some finite subset. For a detailed presentation of the method, see Bishop [18].

11.8 Conclusions

For more information on simulations, tutorials and programs, the reader should consult http://alps.comp-phys.org. The ALPS (Algorithms and Libraries for Physics Simulations) project provides libraries to facilitate quantum lattices model simulations. A list of Internet addresses for numerical programs is given by Lauchli [6].

References

[1] Landau D P and Binder K 2000 *A Guide to Monte Carlo Simulations in Statistical Physics* (Cambridge: Cambridge University Press)

[2] Wolff U 1989 Collective Monte Carlo updating for spin systems *Phys. Rev. Lett.* **62** 361–64

[3] Evertz H G 2003 The loop algorithm *Adv. Phys.* **52** 1–66

[4] Wang F and Landau D P 2001 Efficient, multiple-range random walk algorithm to calculate the density of states *Phys. Rev. Lett.* **86** 2050

[5] Sandvik A W 2019 Stochastic series expansion methods *Many-Body Methods for Real Materials Modeling and Simulations* ed E Pavarini, E Koch, S Zhang and J Verlag des Forschungszentrum Julich

[6] Lauchli A M 2011 Numerical simulations of frustrated systems *Introduction to Frustrated Magnetism* ed C Lacroix, P Mendels and F Mila (Berlin: Springer) p 481

[7] Plischke M and Bergersen B 2003 *Equilibrium Statistical Physics* (Singapore: World Scientific)

[8] Oitmaa J, Hamer C and Zheng W 2006 *Series Expansion Methods for Strongly Interacting Lattice Models* (Cambridge: Cambridge University Press)

[9] White S R 1992 Density matrix formulation for quantum renormalization groups *Phys. Rev. Lett.* **69** 2863–66
[10] White S R 1993 Density-matrix algorithms for quantum renormalization groups *Phys. Rev.* B **48** 10345–56
[11] Schollwock U 2005 The density-matrix renormalization group *Rev. Mod. Phys.* **77** 259–315
[12] Kadanoff L P 1966 Scaling laws for Ising models near T_C *Fizika* **2** 263–72
[13] Wilson K G and Kogut J B 1974 The renormalization group and the ε expansion *Phys. Rep.* **12** 75–199
[14] Tezuka M 2007 An improved initialization procedure for density-matrix renormalization group *J. Phys. Soc. Jpn.* **76** 053001
[15] Troyer M 2015 Computational Quantum Physics http://edu.itp.phys.ethz.ch/fs09/cqp/Script1.pdf
[16] Giannozzi P 2020 Numerical Methods in Quantum Mechanics www.fisica.uniud.it/~giannozz/Corsi/MQ/LectureNotes/mq.pdf
[17] Weisse A and Fehske H 2008 Exact diagonalization techniques *Computational Many-Particle Physics* (*Lecture Notes in Physics* vol 739) ed H Fehshe, R Scneider and A Weisse (Springer) pp 529–44
[18] Bishop R F 1998 The coupled cluster method *Microscopic Quantum Many-body Theories and their Applications* (*Lecture Notes in Physics* vol 510) ed J Navarro and A Polls (Berlin: Springer)pp 1–70

IOP Publishing

Theoretical Tools for Spin Models in Magnetic Systems

Antonio Sergio Teixeira Pires

Appendix A

Group theory

A group G is a set of elements g_1, g_2, …, and a rule for combining (multiplying) any two elements to form a 'product', subject to the following properties:

1. The product $g_i g_j$ of any two group elements must be a group element.
2. Group multiplication is associative: $(g_i g_j)g_k = g_i(g_j g_k)$.
3. There is a unique group element $g_I = I$, called the *identity*, such that $Ig_I = g_I I = g_I$ for all $g_i \in G$.
4. Each element has a unique inverse; that is, for each g_i there is a unique group element g_i^{-1} such that $g_i g_i^{-1} = g_i^{-1} g_i = I$.

A *finite group* is a group with a finite number of elements. If $g_i g_j = g_j g_i$ for every pair of elements in the group, it is called an *Abelian group*.

The group $\mathbb{Z}_2$ is a group with two elements e and x such that $ex = xe = x$ and $e^2 = x^2 = e$. An example is the multiplicative group comprising 1 and −1. This group is used in the study of the Ising model.

A *representation* of a group is a set of square nonsingular matrices (one matrix for each group element), which multiply just like the group elements, and therefore, also constitute a group.

We will be interested in groups with an infinite number of elements, whose elements may be labeled by real parameters, which vary continuously. One example is the group of rotation in three dimensions.

A generating set of a group is a subset such that every element of the group can be expressed as a combination (under the group operations) of finitely many elements of the subset and their inverses.

The group of all unitary $N \times N$ matrices is designed by U(N). In the case, $N = 1$ the group U(1) consists of all complex numbers with absolute value 1. The group of all unitary $N \times N$ matrices with determinant 1 is called SU(N). The group operation is the standard matrices multiplication. A unitary $N \times N$ matrix with determinant 1 has $N^2 - 1$ independent elements, so the group SU(N) has $N^2 - 1$ generators.

doi:10.1088/978-0-7503-3879-0ch12

It is well known that a spinor wave function for the spin ½ particles is described by state vectors in a complex two-dimensional space, and rotation in this space is described by the SU(2) matrices. The generators can be taken as the 3 Pauli matrices. Representations of the SU(2) group by higher-order matrices describe higher values of spins. We show below representations for spins 1, 3/2 and 2. Note that the 3 × 3 matrices for spin-1 do not form a SU(3) group.

In the following, we show some representations for the group SU(2):

For $S = ½$

$$S^x = \frac{1}{2}\begin{pmatrix} 0 & 1 \\ 1 & 0 \end{pmatrix}, \quad S^y = \frac{1}{2}\begin{pmatrix} 0 & -i \\ i & 0 \end{pmatrix}, \quad S^z = \frac{1}{2}\begin{pmatrix} 1 & 0 \\ 0 & -1 \end{pmatrix}.$$

For $S = 1$

$$S^x = \frac{1}{\sqrt{2}}\begin{pmatrix} 0 & 1 & 0 \\ 1 & 0 & 1 \\ 0 & 1 & 0 \end{pmatrix}, \quad S^y = \frac{1}{\sqrt{2}i}\begin{pmatrix} 0 & 1 & 0 \\ -1 & 0 & 1 \\ 0 & -1 & 0 \end{pmatrix}, \quad S^z = \begin{pmatrix} 1 & 0 & 0 \\ 0 & 0 & 0 \\ 0 & 0 & -1 \end{pmatrix}.$$

For $S = 3/2$

$$S^x = \frac{1}{2}\begin{pmatrix} 0 & \sqrt{3} & 0 & 0 \\ \sqrt{3} & 0 & 2 & 0 \\ 0 & 2 & 0 & \sqrt{3} \\ 0 & 0 & \sqrt{3} & 0 \end{pmatrix}, \quad S^y = \frac{1}{2i}\begin{pmatrix} 0 & \sqrt{3} & 0 & 0 \\ -\sqrt{3} & 0 & 2 & 0 \\ 0 & -2 & 0 & \sqrt{3} \\ 0 & 0 & -\sqrt{3} & 0 \end{pmatrix},$$

$$S^z = \begin{pmatrix} 3/2 & 0 & 0 & 0 \\ 0 & 1/2 & 0 & 0 \\ 0 & 0 & -1/2 & 0 \\ 0 & 0 & 0 & -3/2 \end{pmatrix}.$$

For $S = 2$

$$S^x = \frac{1}{2}\begin{pmatrix} 0 & 2 & 0 & 0 & 0 \\ 2 & 0 & \sqrt{6} & 0 & 0 \\ 0 & \sqrt{6} & 0 & \sqrt{6} & 0 \\ 0 & 0 & \sqrt{6} & 0 & 2 \\ 0 & 0 & 0 & 2 & 0 \end{pmatrix}, \quad S^y = \frac{1}{2i}\begin{pmatrix} 0 & 2 & 0 & 0 & 0 \\ -2 & 0 & \sqrt{6} & 0 & 0 \\ 0 & -\sqrt{6} & 0 & \sqrt{6} & 0 \\ 0 & 0 & -\sqrt{6} & 0 & 2 \\ 0 & 0 & 0 & -2 & 0 \end{pmatrix},$$

$$S^z = \frac{1}{2}\begin{pmatrix} 2 & 0 & 0 & 0 & 0 \\ 0 & 1 & 0 & 0 & 0 \\ 0 & 0 & 0 & 0 & 0 \\ 0 & 0 & 0 & -1 & 0 \\ 0 & 0 & 0 & 0 & -2 \end{pmatrix}.$$

The group SU(3) has eight generators that are taken as the Gell-Mann matrices. Now, when expressing all three spin matrices and all five quadrupolar operators in terms of the basis (6.19), we find that the spin matrices become equal to the three imaginary Gell-Mann matrices. In contrast, the five quadrupolar operators equal the remaining five real Gell-Mann matrices.

The *symplectic* group of $2N \times 2N$ matrices Sp(N) is the set of unitary matrices U, which satisfy the condition $U^t \mathfrak{J} U = \mathfrak{J}$ where

$$\mathfrak{J} = \begin{pmatrix} 0 & 1 & & & & \\ -1 & 0 & & & & \\ & & \bullet & & & \\ & & & \bullet & & \\ & & & & 0 & 1 \\ & & & & -1 & 0 \end{pmatrix}.$$

A.1 Gauge theory

Wen [1] defines a gauge theory as a theory where we use more than one label to label the same quantum state. The idea is a generalization from electromagnetism, where a gauge change in the potential vector does not change the electromagnetic fields. A global U(1) transformation is given by $\phi \rightarrow e^{i\alpha}\phi$, where α is a constant. If $\alpha(\vec{r}, t)$ we say that we have a local U(1) transformation. A U(1) gauge theory is a theory invariant under a U(1) transformation.

Reference

[1] Wen X 2004 *Quantum Field Theory of Many-Body Systems* (Oxford: Oxford University Press)

IOP Publishing

Theoretical Tools for Spin Models in Magnetic Systems

Antonio Sergio Teixeira Pires

Appendix B

Green function in magnetic models

In this appendix, I present the method of perturbation theory for the Green function in a way that is more adequate to be used to treat magnetic models. In general, as we saw in chapter 4, the interacting Hamiltonian is of the form

$$H_I = \frac{1}{2}\sum_{q,l,m,n} A(l,\, q,\, m,\, n)a_l^\dagger a_q^\dagger a_m a_n, \tag{B.1}$$

where $l,\ q,\ m,\ n$ here are momentum variables. As we saw in chapter 8, the first term in equation (8.37) is the Green function for noninteracting particles (here magnons). The second term is given by

$$\begin{aligned} \delta G^{(1)}(k_2,\, k_1,\, t_2 - t_1) = &-\frac{1}{2}\sum_{q,l,m,n} A(q,\, l,\, m,\, n) \\ &\int_{-\infty}^{\infty} dt_1' \langle 0|T\{a_l^\dagger(t_1')a_q^\dagger(t_1')a_m(t_1')a_n(t_1')a_{k_2}(t_2)a_{k_1}^\dagger(t_1)\}|0\rangle. \end{aligned} \tag{B.2}$$

We see that the first two terms in the integrand in the above equation vanish, i.e.,

$$\begin{aligned} &\langle 0|Ta_l^\dagger(t_1')a_q^\dagger(t_1')|0\rangle\langle 0|Ta_m(t_1')a_n(t_1')|0\rangle\langle 0|Ta_{k_2}(t_2)a_{k_1}^\dagger(t_1)|0\rangle = 0 \\ &\langle 0|Ta_l^\dagger(t_1')a_m(t_1')|0\rangle\langle 0|Ta_n(t_1')a_{k_2}(t_2)|0\rangle\langle 0|Ta_q^\dagger(t_1')a_{k_1}^\dagger(t_1)|0\rangle = 0. \end{aligned} \tag{B.3}$$

The non-null terms are

$$\begin{aligned} &\langle 0|Ta_l^\dagger(t_1')a_m(t_1')|0\rangle\langle 0|Ta_q^\dagger(t_1')a_n(t_1')|0\rangle\langle 0|Ta_{k_2}(t_2)a_{k_1}^\dagger(t_1)|0\rangle \\ &\langle 0|Ta_l^\dagger(t_1')a_n(t_1')|0\rangle\langle 0|Ta_q^\dagger(t_1')a_m(t_1')|0\rangle\langle 0|Ta_{k_2}(t_2)a_{k_1}^\dagger(t_1)|0\rangle \\ &\langle 0|Ta_l^\dagger(t_1')a_{k_2}(t_2)|0\rangle\langle 0|Ta_q^\dagger(t_1')a_m(t_1')|0\rangle\langle 0|Ta_n(t_1')a_{k_1}^\dagger(t_1)|0\rangle \\ &\langle 0|Ta_l^\dagger(t_1')a_m(t_1')|0\rangle\langle 0|Ta_q^\dagger(t_1')a_{k_2}(t_2)|0\rangle\langle 0|Ta_n(t_1')a_{k_1}^\dagger(t_1)|0\rangle \\ &\langle 0|Ta_l^\dagger(t_1')a_n(t_1')|0\rangle\langle 0|Ta_q^\dagger(t_1')a_{k_2}(t_2)|0\rangle\langle 0|Ta_m(t_1')a_{k_1}^\dagger(t_1)|0\rangle \\ &\langle 0|Ta_l^\dagger(t_1')a_{k_2}(t_2)|0\rangle\langle 0|Ta_q^\dagger(t_1')a_n(t_1')|0\rangle\langle 0|Ta_m(t_1')a_{k_1}^\dagger(t_1)|0\rangle. \end{aligned} \tag{B.4}$$

doi:10.1088/978-0-7503-3879-0ch13

Using Wick's theorem, we write:

$$\begin{aligned}\delta G^{(1)}(k_2, k_1, t_2 - t_1) = \frac{i}{2} \sum_{q,l,m,n} A(q, l, m, n) \\ \int_{-\infty}^{\infty} dt_1' \Big\{ G^{(0)}(m, l, t_1' - t_1')G^{(0)}(n, q, t_1' - t_1')G^{(0)}(k_2, k_1, t_2 - t_1) \\ + G^{(0)}(n, l, t_1' - t_1')G^{(0)}(m, q, t_1' - t_1')G^{(0)}(k_2, k_1, t_2 - t_1) \\ + G^{(0)}(k_2, l, t_2 - t_1')G^{(0)}(m, q, t_1' - t_1')G^{(0)}(n, k, t_1' - t_1) \\ + G^{(0)}(m, l, t_1' - t_1')G^{(0)}(k_2, q, t_2 - t_1')G^{(0)}(n, k_1, t_1' - t_1) \\ + G^{(0)}(n, l, t_1' - t_1')G^{(0)}(k_2, q, t_2 - t_1')G^{(0)}(m, k_1, t_1' - t_1) \\ + G^{(0)}(k_2, l, t_2 - t_1')G^{(0)}(n, q, t_1' - t_1')G^{(0)}(m, k_1, t_1' - t_1) \Big\}.\end{aligned} \tag{B.5}$$

Using the following relation for noninteracting particles

$$G^{(0)}(k, q, t) = G^{(0)}(k, t)\delta_{kq}, \tag{B.6}$$

we can write:

$$\begin{aligned}&\text{First term: } G^{(0)}(m, 0)G^{(0)}(n, 0)G^{(0)}(k_1, t_2 - t_1)\delta_{lm}\delta_{nq}\delta_{k_1k_2} \\ &\text{Second term: } G^{(0)}(n, 0)G^{(0)}(m, 0)G^{(0)}(k_1, t_2 - t_1)\delta_{nl}\delta_{mq}\delta_{k_1k_2} \\ &\text{Third term: } G^{(0)}(k_2, t_2 - t_1')G^{(0)}(m, 0)G^{(0)}(k_1, t_1' - t_1)\delta_{k_2l}\delta_{mq}\delta_{nk_1} \\ &\text{Fourth term: } G^{(0)}(k_2, t_2 - t_1')G^{(0)}(m, 0)G^{(0)}(k_1, t_1' - t_1)\delta_{nk_1}\delta_{ml}\delta_{k_2q} \\ &\text{Fifth term: } G^{(0)}(k_2, t_2 - t_1')G^{(0)}(n, 0)G^{(0)}(k_1, t_1' - t_1)\delta_{k_2q}\delta_{nl}\delta_{mk_1} \\ &\text{Sixth term: } G^{(0)}(k_2, t_2 - t_1')G^{(0)}(n, 0)G^{(0)}(k_1, t_1' - t_1)\delta_{kl}\delta_{nq}\delta_{mk_1}.\end{aligned} \tag{B.7}$$

The first and second terms are represented by disconnected diagrams and we neglect them. In general, the Hamiltonian imposes restrictions in the momentum l, q, m, n such that there is momentum conservation in each vertex. These restrictions appear as δ functions in the factor $A(l,q,m,n)$. For instance, we could have a term such as

$$H_I = \frac{1}{2}\sum_{kqm} A(k - q, q, m, k - m)a_q a_{k-q} a_m^\dagger a_{k-m}^\dagger. \tag{B.8}$$

In the third term of (B.7) we have $k_2 = k - q$, $m = q$, $k - m = k_1$ and so $k_2 = k_1$. Each term from the third to the sixth contributes to the same factor

$$\delta G^{(1)}(k_2, t_2 - t_1) = 2i\sum_q A(q)\int_{-\infty}^{\infty} dt_1' G^{(0)}(k_2, t_2 - t_1')G^{(0)}(q, 0)G^{(0)}(k_2, t_1' - t_1), \tag{B.9}$$

where, for simplicity, I have written the A term as $A(q)$. Introducing the Fourier transform

$$G^{(0)}(k, t_1 - t_2) = \int \frac{d\omega}{2\pi} G^{(0)}(k, \omega)e^{i\omega(t_1 - t_2)}, \tag{B.10}$$

we can write

$$\delta G^{(1)}(k_2, k_1, t_2 - t_1) = 2\delta_{k_2 k_1} i \sum_q \int \frac{d\omega}{2\pi} e^{i\omega(t_2 - t_1)} G^{(0)}(k_2, \omega) G^{(0)}(q, t = 0) G^{(0)}(k_2, \omega) \quad \text{(B.11)}$$

or

$$\delta G^{(1)}(k_2, k_1, t_2 - t_1) = 2\delta_{k_2 k_1} \int_{-\infty}^{\infty} \frac{d\omega}{2\pi} e^{i\omega(t_2 - t_1)} \delta G^{(1)}(k_2, \omega), \quad \text{(B.12)}$$

where

$$\delta G^{(1)}(k, \omega) = i \sum_q A(q) G^{(0)}(k, \omega) G^{(0)}(q, t = 0) G^{(0)}(k, \omega). \quad \text{(B.13)}$$

Note that

$$\lim_{t \to 0^-} G(0)(q, t) = \lim_{t \to 0^-} \int_{-\infty}^{\infty} \frac{d\omega}{2\pi} G^{(0)}(q, \omega) e^{-i\omega t}. \quad \text{(B.14)}$$

Then we can write (B.13) as

$$\delta G^{(1)}(k, \omega) = [G^{(0)}(k, \omega)]^2 i \sum_q A(q) G^0(k, t = 0). \quad \text{(B.15)}$$

For the second-order term, we have

$$\begin{aligned} \delta G^{(2)}(k_2, k_1, t_2 - t_1) = \frac{i}{2!} \int dt_1' dt_2' \langle 0 | T \Bigg\{ & \sum_{kqm} A(k, q, m) a_q(t_1') a_{k-q}(t_1') a_m^{\dagger}(t_1') a_{k-m}^{\dagger}(t_1') \\ & \sum_{ijl} A(i, j, l) a_j(t_2') a_{i-j}(t_2') a_l^{\dagger}(t_2') a_{i-l}^{\dagger}(t_2') a_{k_2}(t_2) a_{k_1}^{\dagger}(t_1) \Bigg\} | 0 \rangle. \end{aligned} \quad \text{(B.16)}$$

Using Wick's theorem, let us consider the term

$$\begin{aligned} & \frac{(-1)}{2!} \sum_{kqm} \sum_{ijl} A(k, q, m) A(i, j, l) \int dt_1' dt_2' G^{(0)}(q, k_1, t_1' - t_1) G^{(0)}(k_2, l, t_2 - t_2') \\ & G^{(0)}(k - q, i - l, t_1' - t_2') G^{(0)}(j, m, t_2' - t_1') G^{(0)}(i - j, k - m, t_2' - t_1'). \end{aligned} \quad \text{(B.17)}$$

We have $q = k_1$, $l = k_2$, $k-q = i-l$, $j = m$, $i - j = k - m$, which gives $i = k$ and $k_2 = k_1$. Hence, we can write

$$\begin{aligned} & \frac{(-1)}{2!} \sum_{k,m} A(k, k_1, m) A(k, m, k_2) \int dt_1' dt_2' G^{(0)}(k_1, t_1' - t_1) G^{(0)}(k_2, t_2 - t_2') \\ & G^{(0)}(k - k_1, t_1' - t_2') G^{(0)}(m, t_2' - t_1') G^{(0)}(k - m, t_2' - t_1'). \end{aligned} \quad \text{(B.18)}$$

Using equation (B.9) we can write the above expressions as

$$\frac{(-1)}{2!}\sum_{k,m} A(k,\, k_1,\, m)A(k,\, m,\, k_2)\frac{1}{(2\pi)^5}$$
$$\int dt_1' dt_2' \left\{ \int d\omega_1 d\omega_2 d\omega_3 d\omega_4 d\omega_5 G^{(0)}(k_1,\, \omega_1)G^{(0)}(k_1,\, \omega_2) \right. \tag{B.19}$$
$$\left. G^{(0)}(k - k_1,\, \omega_3)G^{(0)}(m,\, \omega_4)G^{(0)}(k - m,\, \omega_5)e^{i\omega_1(t_1' - t_1)}e^{i\omega_2(t_2' - t_2)}e^{i\omega_3(t_1' - t_2')}e^{i\omega_4(t_2' - t_1')}e^{i\omega_5(t_2' - t_1')} \right\}.$$

We have the following results

$$\begin{aligned} &\int dt_1' e^{it_1'(\omega_1 + \omega_3 - \omega_4 - \omega_5)} \rightarrow \omega_1 + \omega_3 = \omega_4 + \omega_5 \\ &\int dt_2' e^{it_2'(-\omega_2 - \omega_3 + \omega_4 + \omega_5)} \rightarrow \omega_2 + \omega_3 = \omega_4 + \omega_5, \end{aligned} \tag{B.20}$$

which lead to $\omega_1 = \omega_2$. Equation (B.19) can then be written as

$$\begin{aligned} &\frac{(-1)}{2!}\sum_{k,m} A(k,\, k_1,\, m)A(k,\, m,\, k_2)\frac{1}{(2\pi)^3} \\ &\int d\omega_1 d\omega_3 d\omega_4 G^{(0)}(k_1,\, \omega_1)G^{(0)}(k_1,\, \omega_1)G^{(0)}(k - k_1,\, \omega_3) \\ &G^{(0)}(m,\, \omega_4)G^{(0)}(k - m,\, \omega_1 + \omega_3 - \omega_4)e^{i\omega_1(t_2 - t_1)}. \end{aligned} \tag{B.21}$$

Let $\omega_3 = \omega' - \omega_1 \rightarrow \omega_5 = \omega' - \omega_4$. Using this result in equation (B.20) we write

$$\frac{1}{2\pi}\int d\omega_1 e^{i\omega_1(t_2 - t_1)} \left\{ \frac{(-1)}{2!}\sum_{k,m} A(k,\, k_1,\, m)A(k,\, m,\, k_2)\frac{1}{(2\pi)^2}[G^{(0)}(k_1,\, \omega_1)]^2 \right. \tag{B.22}$$
$$\left. \int d\omega_4 d\omega' G^{(0)}(k - k_1,\, \omega' - \omega_1)G^{(0)}(m,\, \omega_4)G^{(0)}(k - m,\, \omega' - \omega_4) \right\}.$$

The Fourier transform is given by

$$\delta\tilde{G}^{(2)}(k_1,\, t_2 - t_1) = \frac{1}{2\pi}\int d\omega_1 \delta\tilde{G}^{(2)}(k_1,\, \omega_1). \tag{B.23}$$

As we saw in section (8.4), we can cancel the term $1/m!$. We also have two diagrams equivalent to (B.23). Hence, we can write the final result as

$$\begin{aligned} &\delta G^{(2)}(k,\, \omega) = -2\sum_{q,m} A(q,\, k,\, m)A(q,\, m,\, k)[G^{(0)}(k,\, \omega)]^2 \\ &\frac{1}{(2\pi)^2}\int d\omega' d\omega'' G^{(0)}(q - k,\, \omega' - \omega)G^{(0)}(m,\, \omega'')G^{(0)}(q - m,\, \omega' - \omega''). \end{aligned} \tag{B.24}$$

www.ingramcontent.com/pod-product-compliance
Ingram Content Group UK Ltd.
Pitfield, Milton Keynes, MK11 3LW, UK
UKHW051855100926
8556IPUK00005B/635

9 780750 338776